THE BRAINPHONE PROPHECY

Stop Corporations and the Government from Inserting a Smartphone in Your Brain

Scott Snair, Ph.D.

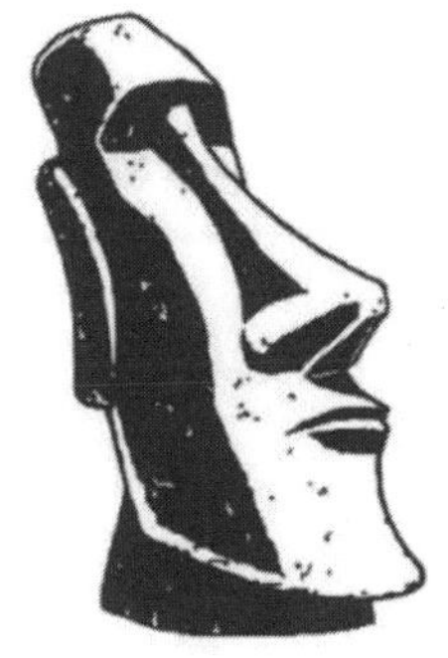

Adventures Unlimited Press

The Brainphone Prophecy

ISBN 978-1-948803-48-9

Published by:
Adventures Unlimited Press
One Adventure Place
Kempton, Illinois 60946 USA
auphq@frontiernet.net

Cover design and illustrations by A.J. Snair

AdventuresUnlimitedPress.com

10 9 8 7 6 5 4 3 2 1

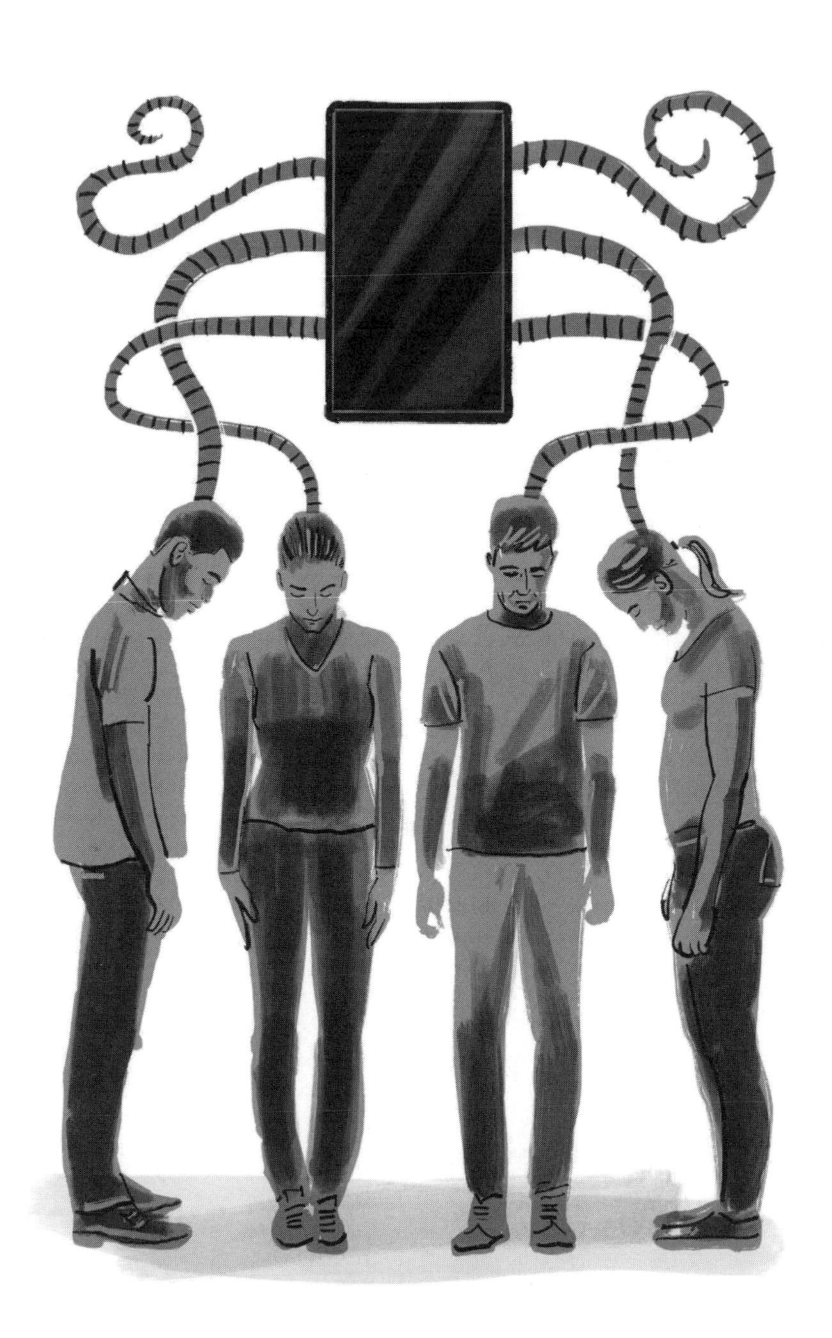

This book is for Jack,
and for the brave new world he's about to enter.

THE BRAINPHONE PROPHECY

Stop Corporations and the Government from Inserting a Smartphone in Your Brain

Scott Snair, Ph.D.

Adventures Unlimited Press

TABLE OF CONTENTS

ACKNOWLEDGMENTS

Many thanks to the good people who supported this project. A special thank you to my agent, Anne Devlin, who saw merit in warning people about the upcoming brainphone. Without her enthusiasm, I never would have completed this book. Thank you to the great folks at Adventures Unlimited Press for taking on this title and promoting it. They include David Hatcher Childress (publisher), Jennifer Bolm (editor), and Lori Brandner (manager).

I appreciate Duncan Roads, the editor of *Nexus Magazine*, who initially published my lengthy cover story on the brainphone. He understands what's about to happen better than anyone I know.

I also appreciate A.J. Snair, the full-time commercial illustrator who designed the cover and did all the amazing illustrations for this book. Brother, we've waited a long time to work together. I'm glad we finally had the opportunity to do so.

Thanks to Bob Eisiminger, who got me interested in this topic by hiring me into his cybersecurity company as an old guy. (I meant me, not him—although we're both the same age.) His mentorship has meant a lot. I'm also grateful to my work colleagues, Kevin Marbray (team leader) and Ariel Sledz (teammate).

Many thanks to Brian and Lisa Moss, and Gary and Kate McCabe, for their advice and friendship.

A special thank you to my wife, Mary-Jane Snair, Ph.D., for her edits and suggestions. Mary-Jane, I love you with all my heart.

Finally, I love and appreciate my daughters, Patti and Katie, who continue to love, tolerate, and encourage me.

INTRODUCTION: MEETING MY FIRST HIGH-TECH BILLIONAIRE

It is easier for a camel to go through the eye of a needle than for a rich man to enter the kingdom of God.
(Mark 10: 25 NKJV)

Not too long ago, I met my first high-tech billionaire and spoke with him for about two hours at one of his homes. It was a job interview: I was meeting him about the possibility of running one of his business ventures. (Per a non-disclosure agreement, I am not identifying the person or the position.)

As I arrived, I said hello into the intercom, and the gate of his estate opened remotely. When I passed through the entrance, it occurred to me that my car was possibly the oldest, dumpiest automobile ever to journey onto those grounds.

I went to the building I had been instructed to find. I saw him there, by himself, drywalling his pool house. I didn't do much talking, if any talking at all. What I did do for the next two hours was listen to him as he talked about himself. He spoke of his wealth and his influence. He listed his enemies. He discussed the federal tax auditors who were trying to take his money. He elaborated, in brutal detail, about the lawsuits of those who, like the auditors, were trying to take his money.

He also spoke of his family, particularly a son who had rejected the family enterprise and the many trappings of wealth and had left him to become a bartender. "I suppose I shouldn't be upset," he suggested. "Bartenders don't do too badly. After all, he probably makes about double what I'm going to offer you!" I laughed at the joke. Only it wasn't a joke. A week later, I was offered the job: the salary was, indeed, about half of what most bartenders make and less than I had made in about 30 years. I politely turned down the job and took another one that paid three times what he had offered me. I have no doubt someone else took the position—someone who, for whatever reasons, questioned his or her own self-worth and accepted the role of minion.

As I drove home from my meeting with this billionaire, it dawned on me that this man was nothing like me or anyone I had ever met before. Most people I know don't worry much about money, because they don't have any money to worry about! When meeting someone for the first time—even when conducting a job interview—most people might talk about their business, their community, their hobbies, what they do when they get away for the weekend, or how much they love or hate the local weather. They also might show an interest, even as only a common courtesy, in the person they are talking to. In contrast, this man seemed lonely, bitter, paranoid, obsessive, and way, way, way too self-focused. The encounter truly shook me.

Furthermore, I realized that adding a substantial amount of power to these types of personality traits could not be a good thing for humankind, particularly as the extremely wealthy become wealthier and more influential and the rest of us are forced to have less. Does being super rich make you crazy, mean, and conniving? I don't know. But I always have thought that the mundane money troubles most of us endure have a tendency to ground us mentally and to put the good things about life into perspective. Perhaps this lack of grounding is what makes rich people the odd characters they are.

An interesting end to my visit: As I thanked this high-tech billionaire for his time, and as he walked me to the door, we passed a brand-new Samsung Smart TV, mounted on a wall. "I just bought it, and I'm already thinking of getting rid of it," he remarked, gesturing towards the television.

"Why is that, sir?" I asked.

"Because I just found out that the voice detection is always on, even when the TV is off, and that what I say is heard and collected by the company."

"Wow, what a violation of privacy!"

"The thing that surprises me most," he continued, "is no one seems to be all that outraged. People seem relatively uncaring about a corporation listening in and keeping track of what we say in our homes all day long." He stared for a moment at the television—almost as if he were upset that he hadn't thought of

Perhaps not being grounded by daily, mundane money problems is what makes rich people the odd characters they are. Photograph by Anastase Maragos. Courtesy of the photographer via Unsplash.

the idea himself.

Maybe what he was thinking was, "I need to get busy!"

Do you think your modern technology might be monitoring too much of your life these days? Wait till they insert it in your brain!

Due to a perfect storm between a rich, controlling group of corporations and futuristic phone technology that mesmerizes us, we soon will be asked to have a next generation of smartphones physically inserted in our brains. Most of us will say yes, buying into the wonders the technology promises. Several giants in the business world of informational technology already are discussing what life will be like when man and machine are merged. These tech giants are excited. The rest of us should be terrified. The reality is likely to be nightmarish, and the transformation will forever change who we are as living beings.

This book argues that we are about to enter a new phase of human existence, as we are merged with a technology that monitors us, alters us, and, in many ways, commands us.

Fortunately, there are ways we can resist this fusion, if we choose to push back. This book tells how.

Is this work the manifesto of someone who is anti-technology or anti-capitalism? No. Not even close. On the technology side, I am a data analyst for a major cybersecurity company. I was once the director of technology for a school, and I have taught the use of educational software and quantitative analytical software to students—many of them doctoral students—for many, many years. On the business side, I am a pro-business (albeit moderate) Republican voter in the United States. I believe in capitalism and the amazing benefits of a free market economy. I believe in the American Dream—heck, I'm living it.

However, my background also includes time in the U.S. Army, where I saw firsthand in other parts of the world how quickly people lose control of their lives when they let their guard down. My military experience carried into my civilian work experience, including full-time work for a couple of years at Rutgers Business School, where I helped returning combat veterans write business plans and gain loans to start up their own businesses in New Jersey. I was also a university dean for some years, mentoring students as they pursued their doctorates in counterterrorism and national security. It was in these capacities that I immersed myself, research-wise, in what actually threatens the United States and the quality of life many of us enjoy. Are radical Islamic terrorists—the thing we have spent trillions of dollars protecting ourselves from—really the thing that challenges America the most? I don't think so. The truth is that I'm not sure they even belong in the top five. Way more menacing to this country are the opioid epidemic, the growing national debt, security issues related to global warming, the growing dominance of corporations, and an unhealthy craving for technology so strong that, soon, we are likely to allow devices to be inserted in our brains.

This book suggests that nothing is more threatening to the lives of free people than the merging of these last two items—the growing power of a few very big companies, and technology placed into our bodies that can monitor us and eventually control us. It also argues that this historically monumental, upcoming event is not an accident. It is by design. Corporations that hold sway over technology—encouraged by governments obsessed with control—have it out for us, and their plan threatens not only the quality of our lives, but also our health, our desire for peace and self-determination, and, really, our very existence as humans (or, at least, how *human* is currently defined).

This narrative lays out how we're being played. It touches on the possibility that the upcoming merging of man and machine might have been forecasted by prophets in *The Holy Bible*. It offers how we can identify the tactics of those who would like to insert technology in us, and how we should react to these tactics. It suggests some ways we can push back and protect ourselves.

Keep in mind that the next generation of smartphones, for insertion directly in the human brain, will be introduced in ways that are painless and hypnotic—like the Samsung televisions that hear our voices and seem to follow our directions. It might happen without much fanfare or controversy. People initially might love the convenience. Corporations (and governments) certainly will love being tapped into everyone's thoughts and inclinations. But the transformation ultimately will be horrifying. We shouldn't allow this hideous morphing of man and machine to happen. This book explains how we might stop it.

PART ONE:

THE INTENT TO COMMANDEER US

CHAPTER 1: WHY ARE YOU ABOUT TO HAVE A SMARTPHONE INSERTED IN YOUR SKULL?

You are the children of the Lord your God: you shall not cut yourselves nor shave the front of your head for the dead. (Deuteronomy 14: 1 NKJV)

You Will Feel the Social Pressure for Getting the Implant

One day soon, the latest edition of your smartphone will include a very special protective case—your skull. Why? Because people will be standing in line to have their newly purchased devices implanted in their brains.

That's right. A few, short years from now, you will be given the option of having a smartphone-like device inserted in your brain. The social pressure for doing so will be strong, as many of your friends will be having it done, and you will not want to feel left out. Your employer might require it. Having a brainphone will allow you to do all sorts of things: call up the Web without needing a screen; talk to your friends through your thoughts; have a second language, such as French or German, downloaded into your brain where you immediately will understand and speak it; play a video game while immersed in a life-like virtual world; or have your addiction or anxiety instantly cured. Cool, right?

Hmmm, maybe not so much. Unfortunately, the flip side to owning a brainphone is likely to be horrific. You always will be connected to the Web, to social media, and to the thoughts of others—even at night while you sleep. You always will be collectively judged by your family, your friends, your professional network, corporations, and, likely, the federal government. To the extent that thoughts and electronic data become intertwined and indistinguishable, your thoughts might be deleted or changed by powerful people and entities who do not like the things you are thinking on your own. You likely will not have your own life: you essentially will be part of a collective.

A new language, such as French or German, will be downloaded in your brain, and you instantly will understand it.

During the Summer of 2019, with very little fanfare—and approximately zero outrage—high-tech billionaire Elon Musk announced plans to begin experimenting on humans with Neuralink devices (also called *links*), brain implants offering all kinds of improvements to how your brain processes information and how it connects to the Web. Some of these promises include learning another language instantly; having Parkinson's disease or Alzheimer's disease treated through brain reprogramming; and having your bad habits cured by washing your brain of the compulsive tendencies that create them (Markoff, 2019).

And so, you and your friends are about to face an important, race-changing decision: Are you so mesmerized by smartphones that you are willing to have them implanted in your brains, forever changing who you are as creatures? Or are you committed to staying who you are and resisting the allure of the technology? Again, it is not a decision for science fiction or for a distant future. Events are happening right now that are bringing the human race to this important, existential crossroads. Do people enthusiastically allow technology to be placed in their

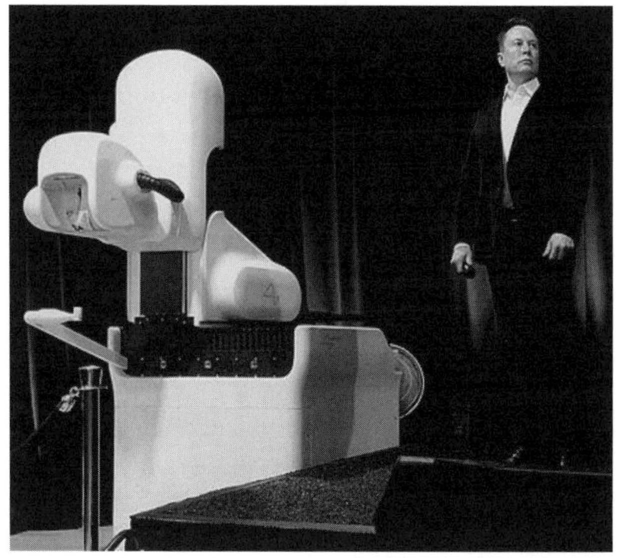

Elon Musk with the robot that performs the link implants. Photograph by Steve Jurvetson. Courtesy of the photographer via Creative Commons. License https://creativecommons.org/licenses/by/2.0/deed.en

brains, making their minds work very differently (perhaps better, perhaps not)? Or do people turn their backs on this tech, perhaps even passing laws banning the merging of processors and brain tissue? Do people draw the line between humanity and machinery?

One could argue that humans have been planning their own extinction for some time. From overpopulating, to ravaging the world's resources, to building nuclear weapons, to warming the earth with greenhouse pollutants, to poorly planning for pandemics—well, people could not have come up with better ways to write themselves off if they had designed them in a boardroom. Both scientists and science fiction writers continue to offer realistic scenarios where Earth is one day sans humans.

However, in most science fiction stories, the person of the future is very much like the person of today: the surroundings are different, but the person is a constant. That is, in the movies, when the characters face extinction, you can relate, because they are the same as you. But perhaps there is a frightening midpoint, where people still exist on Earth but where they are not at all like their former selves. By merging their brains with technology, perhaps people will become an entirely different species in a couple of decades. Perhaps the human race, in its current form, is about to end much earlier than previously thought.

Your Friends Will Say It's Easy

Even the earliest version of brainphone installations is projected to be no more complicated than, say, today's LASIK surgery for correcting your vision. Later on, just as getting your ears pierced used to be a medical procedure at the doctor's office but later became a retail procedure, having a brainphone inserted might be something you do in a pagoda at the mall. The procedure will be fast, painless, convenient, and eventually cheap enough for everyone to afford—making having one all the more tempting. "But, mom," you might overhear someone say, "*all* the kids are getting brainphones!" The first iteration of brainphones, currently in design, is not terribly intrusive. A small piece of hair and skin is cut and peeled back on your head. A round, one-inch

Getting your brainphone inserted might happen in a pagoda at the mall.

piece of your skull is augered and removed for good, replaced by the brainphone, with the wiring interwoven with strands of your brain tissue. Your skin is reattached, and no one can see your new device. With advanced miniaturization and further understanding of the mind, later models will be smaller and even easier to mount.

So, your brainphone is installed and you're ready to go. The device contains a battery, an inductive charger (for recharging the device without having to plug it in), a processor, a Bluetooth (with modems likely in later iterations), and, of course, the electrodes intertwined with your brain's neurons. You now can see the Web in your mind, move a cursor, click, and select menus just by thinking about it, and communicate—through your mind—with other people who have brainphones. (The "talking without words" aspect is, in fact, one of the earliest features revealed by Elon Musk about his Neuralink device.)

You now can drive with an interactive, transparent map overlaid in your vision on the actual highway. You can call and text or talk to your friends without losing sight of the road while you drive. You can take photos and videos and store them just by looking and thinking. If you're in school, you can do calculations in your mind. Better yet, you can ask your desk neighbor for the answers to the pop quiz without talking or moving your eyes away from the quiz—in other words, without your teacher knowing what you're up to. Pretty smooth, huh?

But the real fun is with gaming. You can pull up your favorite games just by thinking about them. The other characters in the game are other people with brainphones, connected via the game's software and the Web and fully existing in your head. The game's setting is a 360-degree, virtual, extremely realistic, alternate world. You can look like anything you want and have any weapon or powers you want (within the game's parameters, of course). Watch out, though—when a competitor hits you with his or her weapon or powers, you will get knocked down, and it will feel real. Depending on the game, it might even hurt.

Back to the real world. The new girl who just moved in next door is very attractive, but she seems to speak primarily Spanish. No problem. You order Spanish via the Web. The language gets inserted via your brainphone directly in your memory, and you're

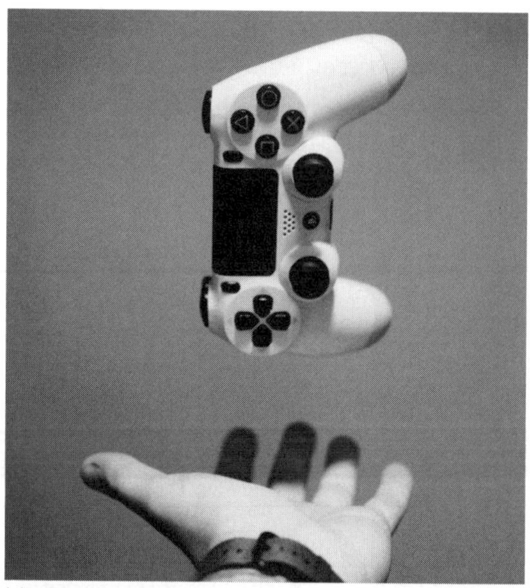

You will be able to pull up your favorite games in your mind just by thinking about them. Photograph by Nikita Kachanovsky. Courtesy of the photographer via Unsplash.

instantly able to speak Spanish and carry on a conversation with her. You also can instantly purchase and download into your mind the knowledge you need to repair your car or knowledge on martial arts to defend yourself (kind of like *Chuck* on that old television show).

Social media will be a whole new world, as you and your friends with brainphones completely immerse yourselves into each other's thoughts, opinions, life happenings, and existences. People without brainphones will feel (and be) completely left out. Too bad for them! Want to take out that new girl next door? If she has a brainphone, too, then your first date could be virtual. You two can see the Eiffel Tower and take a boat ride down the Seine River, cuddling like all the other young lovers in Paris.

Not everything about your brainphone is fun and frivolous. There are some genuinely good, life-changing things about it. For example, if you have epilepsy, a program administered through your brainphone essentially will rewire your brain and rid it of the disorder and the seizures that come with it. A program might be able to find alternate brain pathing to cure blindness and deafness. Older people with brainphones might have their dementia or Alzheimer's cured. Your new brainphone is really something.

Your Employer Might Require It

For the younger generation that never knew life without a laptop or a smartphone, the decision to get a brainphone will be, well, a no-brainer. Anecdotally, the buzz in schools and online over Musk's Neuralink device is already positive, as the notion of being *one* with one's technology sounds very appealing. After all, the physical detachment of today's smartphones is already a hassle: people die every day from texting while driving; they die every day from crossing the street while staring at their smartphone screens instead of watching out for cars; hundreds die every year from taking selfies while standing on the edge of cool cliffs or scenic bridges. Having the technology wired into one's brain will make texting, photographing, and looking things up on the Web infinitely easier while driving, crossing the street, and rock climbing.

Indeed, it is the connectivity of the home computers and smartphones that made them so popular. Although home computers initially were marketed for their processing capabilities and for the owner's ability to look up information on the Web, it was email, America Online instant messaging, and chat rooms that made the computers a sensation. Similarly, although smartphones bring a myriad of apps and capabilities, it is the texts, short-video messaging, and facilitating of social media that makes them so alluring.

Granted, there is also an older, pre-smartphone generation—already mildly wary of technology—that will be hesitant about implants. This generation tends to have a lot of the purchasing power in America. So how will corporations get the older crowd on board? They likely will borrow a few pages from the playbook described in Naomi Klein's (2007) book, *The Shock Doctrine*, which explains how corporations and governments sometime use major tragedies to gain new controls over people. Klein's thesis is compelling (and, incidentally, it's a great read). So, to follow Klein's template, people otherwise hesitant to have a piece of hardware placed in their brain will come around in a time of tragedy. As an example, Americans, who used to be passionate about their civil liberties, accepted all kinds of infringement of their privacies after the terrorist attacks of September 11, 2001. They became comfortable with getting full-body scans at airports, having their personal computer activity monitored by the federal government, and having their travels tracked by corporations and governments via their smartphones. That is, people were willing to turn over a lot of autonomy to other entities, ostensibly to keep them safer—protected, so to speak—and presumably to prevent another horrible cluster of attacks.

The most likely scenario is that, one day, a pandemic similar to polio will hit humanity, crippling hundreds of thousands of people. The brainphone, with its brain-rewiring capability, will be presented as the cure. And, suddenly, millions of people who swore up and down that they never would allow an implant in their brains will do so gladly. The controversy will be gone (if it ever really existed in the first place). From there, corporations will use the "frog in a pot" approach. The old saw refers to the idea that if you toss a frog in a pot of hot water, it will hop out. But if you put a frog in a room-temperature pot of water, it will stay put. Then, you slowly increase the temperature, until, by the time the frog realizes what's happening, it's too late and he's frog stew. Corporations will start by enticing people with convenience-related brainphone applications, such as calling up the Web simply by thinking about it and seeing the "screen" and moving the cursor in one's mind. They gradually will move to the scarier stuff, like monitoring people's thoughts or downloading the desire to purchase things, later on.

Another way corporations will sell you on the idea of a brainphone is to make your life very difficult without one. Imagine in today's world if you tried to work for a company while refusing to answer your emails. You wouldn't last long. Now project that predicament onto a future with brainphones. Your employer might, indeed, require you to have one. The brainphone might be used for security reasons, such as to allow you access into your building and to keep track of you in the event of emergencies such as fires or workplace shooters. It might be the way you virtually meet with your team. It might be the way you check in with your boss and sign your time card. It

Seems real but could be brainphone-simulated. It might be the way your boss wants to meet regularly with the team. Photograph by Beth MacDonald. Courtesy of the photographer via Unsplash.

might be the way you order goods, services, shipping, and business travel. It might be the only way you can purchase a soda in the cafeteria! In sum, your life might become very, very tough at work without the implant.

So, along with the teenage FOMO ("Fear of Missing Out"), there also will be the adult and breadwinner FOMO, and it will be a powerful persuader in having everyone get a brainphone. From there, corporations—and, likely, governments—will connect, direct, and control people. The arrangement will allow for a small number of corporations to become extremely rich and controlling.

Perhaps just as frightening is how this new phenomenon will immediately change the very nature of who you are on both biological and behavioral levels. In most ways, you will no longer be the creature you once were.

But You Will Be Instantly Changed

Admittedly, you're not physically the same as humans from as recent as a few hundred years ago. You're likely taller because of better nutrition and better health. Your teeth are larger than those of even your recent ancestors because of fluoride and how it thickens and strengthens teeth. And, believe it or not, because of evolution, your mouth is likely slightly larger than those of your relatives from just a generation ago, because global warming has created a genetic need for lips that better evaporate moisture and cool down your face. (No kidding!) And so, the word *mutate* to describe how the brainphone might change you is a bit unfair: we're always changing and evolving.

However, as your brainphone begins interacting with the brain functions that control your growth, healing, body-systems monitoring, and physical and mental well-being, you might find yourself changing in very dramatic ways. Furthermore, if your personality worsens substantially, it might alter your face, as your eyes might become wider with paranoia or your mouth bends down with constant sadness or anger. Compared to how you are today, you might become a real physical and mental mess.

There are possibilities that, as you're tapped into your brainphone for long periods of time, you might curl into a ball in the corner of a room, or your face might display all kinds of severe contortions. As people, aware of these conditions, go off somewhere to access their devices, dark rooms full of people contorting in semi-paralysis, as they travel the Web in their minds, could become the new human hives.

Whether the need for belongingness, as a human cultural trait, is a strength or a weakness, its impact will come down like a hammer when our phones go from our hands into our skulls. Always on. Always connected. At first, people will be able to tell what you're thinking by looking into the search history inside your brain. But, as the technology progresses, they'll eventually have full access to your opinions, thoughts, and desires. On the plus side, lying to people will become much more difficult. When you sign a contract, the person across the table instantly will know if you plan on honoring that contract, and you will know it of him—which is nice. But lacking the ability to lie also means lacking the ability to use social tact and diplomacy. When someone asks you, "Does this business suit make me look fat?", you won't be able to politely say that it looks fine. For that matter, the person wearing that suit will be able to know what everyone thinks about it, instantly, as he walks down the street and receives instant thought transmissions from everyone that looks at him. Imagine someone who is not physically attractive being told that very thing, through the collective thoughts of others, every single day of his life. Or imagine someone who has always been beautiful realizing, via mob thought, that she is now getting older and no longer possesses the same level of beauty.

Collective, interconnected thought is not a new concept. In the science-fiction television show *Star Trek: The Next Generation*, the crew of the starship *Enterprise-D* faces The Borg, alien creatures who hijack different species across the universe, assimilating them and bringing them into their collective, hive mind ("Q Who?", 1989). With the addition of each culture, they become a little bit smarter and a little bit more invincible. Their trademark phrase: "Resistance is futile." Like zombies, only with wires coming out of their heads and faces, they move deliberately as they come for you. They are Starfleet's most formidable enemy, and they are the stuff of nightmares.

In some ways, today's technology already has brought us closer to collective thought. Instant access to information and, through social media, the reactions of other people seem to be magnifying our outrage and creating a new, human hypersensitivity. In their book, *The Coddling of the American Mind*, Greg Lukianoff and Jonathan Haidt lament that even on college campuses—places where a debating of the issues and a wide variety of opinions should be celebrated—the culture has become "ideologically uniform," with administrators and faculty expected to toe the line. Lukianoff and Haidt also draw attention to social media as the platform for *callout culture*, where anyone can be instantly and publicly shamed for saying anything, even something clearly well-intentioned but negatively interpreted by the collective. (More on Lukianoff and Haidt in Chapter Two.) We're already at a point where people can lose their jobs because of the rush to judgment from The Hive, even if the immediate outrage is not based on all the facts, or, sometimes, any facts at all. How will collective judgment and collective hypersensitivity be amplified when we become forever tapped into the thoughts of those around us?

An offshoot of *callout culture* is *cancel culture*. Rather than simply group shaming someone online, *cancel culture* goes the extra step of encouraging others to no longer support or do business with that person. For example, if a local business owner is *cancelled*, people within an online group of contacts boycott his or her business, and they persuade others—through social media—to do the same. The shaming might be applied to a celebrity or a large corporation for not toeing the line of what social-media groups consider politically correct or appropriate. A recent example includes comedian Kevin Hart's stepping down from hosting the Oscars after online backlash for homophobic jokes he had made ten years earlier. Another example is posts on social media encouraging patrons of Equinox fitness clubs to give up their gym memberships over company chair Stephen Ross's fundraising for Donald Trump's reelection campaign. Now imagine the intensity of *cancel culture* when everyone is wired into your thoughts. People will not only be able to tell if you support a *cancelled* corporation: they'll be able to tell if you've done business with that company recently. They'll be able to tell if you've recently watched a *cancelled* pop artist. If they see too much in your brain that they don't like, you might find yourself *cancelled* as well, unable to be a part of certain online groups; unable to attend certain real-life clubs; unable to work for certain companies where political correctness is paramount. You will forever be watched and judged.

One symptom of both *callout culture* and *cancel culture* is that, since communication on a topic is so swift, uniform, and accessible to the masses, there is no time for people to take a breath and, as the saying goes, "let cooler heads prevail." There won't be any cool heads at all, as the processors in our brains will be heating things up, both literally and figuratively! We'll always be getting fed talk of new controversies, new conspiracies, and new outrages, and we'll always be troubled by them.

Some forms of philosophy suggest that reality and truth are external to human thought. In other words, if there's a rock on the ground, it exists, whether or not anyone sees it and whether or not anyone thinks it is there. Furthermore, if a group of people decides, say, in a conference, that the rock is not there, their decision doesn't alter the truth—the truth that the rock is there.

But how difficult will it be to seek external truths when everyone is in one another's heads? What will reality be? What will the truth be? Perhaps The Hive will decide, and that will be that. The result, of course, will be disastrous. If everyone collectively decides that a bridge will hold a certain weight—and ostracizes the engineer with a calculator telling them that they're wrong—they will proceed across the inferior bridge and it will collapse. Then, in some bizarre twisting of the events that just happened, the engineer who warned of the disaster somehow might be collectively blamed and prosecuted!

And the World Around You Will Be Instantly Changed

The condition of having everyone wired into one another at all times might change the very definitions of *culture* and *community*, which usually pertain to people with shared attributes and experiences. Rather than these commonalities forming the communities, it may eventually work the other way around, where the communities form the commonalities, and you have to embrace and take on those traits if you want to be accepted into the fraternity. Finding people with your likemindedness and hanging out with them might no longer be a thing. There will be no one like you. You will examine the requirements and hallmarks of the collective, and you will, as the *Rush* song says, "conform or be cast out."

There are undoubtedly hundreds of questions to be asked about the physical and psychological side effects of having a brainphone. Many—if not most—of these questions have to do with a device that might not be properly vetted or that doesn't work. But maybe these are the wrong questions. Perhaps the more appropriate questions have to do with a brainphone that *does* work. Perhaps that is *the* question: Is a brainphone that works exactly as designed *really* a good idea?

An existence with brainphones seems likely to be an existence without individuality, as The Hive decides what's appropriate and shames you if you don't play along. Civil liberties, particularly those involving freedom of privacy and expression, will cease to exist.

Corporations, eager to maintain collective opinion in support of their business plans, will control the dialogue and keep track

Corporations will control the dialogue about brainphones. Photograph by Note Thanun. Courtesy of the photographer via Unsplash.

of where you are and what you are purchasing. Furthermore, many of the companies that produce brainphones will flourish, as they form large monopolies over parts, maintenance, access to information, and big data regarding what billions of people in hives are thinking and doing. And they will gladly—for a fee—provide electronic realities that help their billions of clients to escape. As former Google product philosopher Tristan Harris mentioned on PBS Newshour in 2017, the smartphone "puts a new choice on life's menu that's sweeter than reality." And this cyber-reality is sure to be all the sweeter once it's in your head.

In the early stages of brainphone technology, corporations will be able to access your thoughts and preferences by tracking what websites you call up in your mind and what types of purchases you make (much like they do now by tracking your smartphone activity). But there is a later iteration, where those who run the science will be able to read your actual thoughts. Companies will like that. Governments will *love* it. The ability of an oppressive, federal government to constantly monitor the inner thoughts of its citizens for any signs of revolution—or, really, any type of non-compliance—is a fascist's dream. It is an idea that even George Orwell could not comprehend when he wrote *1984* over 70 years ago.

Speaking of later iterations, an old philosophical question is worth examining: How much of a person can be replaced and have it still be a person? That is, once all of our brains are part of The Hive, and once robots become advanced enough to take over and accelerate our brainphone technology, how much of us will be us, and how much of us will be Artificial Intelligence (AI)? Will there be any "us" left?

There is an argument out there, among AI enthusiasts, that, once robots begin to outpace humans in how they think and advance, humans shouldn't fight their secondary status and should willingly merge with the robots, creating a sort of transhumanism. In discussing his Neuralink device on "The Joe Rogan Experience" in May of 2020, Elon Musk admitted that his device might be a first step in this merge. Instead of fighting self-aware technology, he suggested, we should, "go along for the ride [with AI]."

Is Musk concerned about how his version of the brainphone someday might dehumanize the human race? You'll someday have to ask his son, born in 2020 and named X Æ A-Xii.

CHAPTER 2:
HOW WILL WE GIVE IN TO THE ADDICTIVE NATURE OF TECHNOLOGY?

No temptation has overtaken you except such as is common to man: but God is faithful, who will not allow you to be tempted beyond what you are able, but with the temptation will also make the way of escape, that you may be able to bear it.
(1 Corinthians 10: 13 NKJV)

We Will Go with the Flow

Smartphones inserted—through a simple, medical procedure—directly in our brains? No way! Not in my lifetime! Not *my* brain! That's *crazy*!

Is it?

I once asked a lecture hall of college students, "Who in here would be willing to wear a monitor on their ankle, all the time, that would allow the federal government to track and record your movements and listen in on all of your conversations?" No one raised his or her hand. Even after I changed the scenario to include some type of national-security emergency or natural disaster, no hand went up. "Why aren't you all raising your hands? You already have such a device."

Confused faces.

"Your smartphones."

Some faces understood—others still looked confused. Many—if not most—had no reaction all at: they were too busy doing something on their smartphones to pay attention!

The idea of having some authority—one's boss, a corporation, the police, the federal government, etc.—attach a monitor onto one's ankle and keep tabs on oneself is distasteful to nearly everyone. But when the tracking is combined with the convenience of communication, travel directions, gaming, and entertainment, it not only is acceptable but somehow forgettable. We have become numb to corporations and governments watching us. Cameras tracing our license plates as we travel (and automatically mailing us toll bills or speeding tickets); face-recognition cameras identifying us from large databases as we walk carelessly through downtown areas; stores following which departments we're shopping in, based on information transmitted from our smartphones; people tracking what we look up on the Web and who we communicate with on our computers—they are all things that would have outraged American citizens only a generation or two ago. But we have grown accustomed to them. These intrusions now are just part of daily living. The smartphone quietly has become the ultimate tool for others to watch us, track us, and listen to us. And, considering how difficult it is for most people to put down, it might as well be attached.

Having smartphones inserted in our skulls (literally) only sounds crazy to us if we forget how much we've already allowed people to infiltrate our minds (figuratively). It only sounds crazy to us if we fail to grasp how quickly we have changed our attitudes about the intrusive nature of technology, as well as how rapidly the technology is advancing. In fact, the technology for

People forget that they're constantly being monitored and tracked.
Photograph by Alan Rodriguez. Courtesy of the photographer via Unsplash.

brainphones already exists and is currently being used in other capacities. Aside from needing a few minor advancements in miniaturization and surgical techniques, a streamlined process for mass producing and mass implanting the devices is right around the corner. It's really just a matter of getting us to jump on board.

So why the outraged responses whenever I bring up this eventuality to a group of adults? Why does it sound so *crazy* to others when they hear about it aloud for the first time? Here are some possibilities. First, their reaction is emotional as much as it is judgmental. Our current, imaginary line dividing human flesh and machinery is a nice, thick, well-defined line (for now), and the idea of having that line crossed seems outrageously disrespectful. The idea of being asked—or forced—en masse to have metal, plastic, and circuitry placed in our skulls strikes us on a deep, subliminal level, and it is instantly unnerving. It sounds physically intrusive—even abusive, like physical or sexual assault. Another reason for the immediate rejection of the idea is that it contradicts our notion that the human mind is uncharted territory, full of mysteries and wonderment. If a corporation can attach a miniature smartphone to the neurons of our brain in a way that makes the technology work, then maybe the brain isn't as mysterious as we thought—and that bugs us! Finally, adults generally reject the idea of brainphones, because the thought is insulting to our intelligence: Surely people aren't going to allow such a thing to be done to them, right?

I argue that not only are humans going to allow such a thing, but that they are going to demand it and look forward to it—much the same way that hundreds of people camp out and stand in line at an Apple store waiting for the next version of the iPhone or the Apple Watch to come out. People are going to fear the feeling of being left out if they don't have their brainphones implanted, and they will rush to have it done. The horror of what is happening might not even be all that apparent to the generations who aren't aware of a different reality, much like the iGen youth of today who don't remember what life without smartphones and social media was like. In other words, I suggest that, eventually, people's reaction to the brainphone is going to be the direct opposite of what adult reactions to the *idea* of the brainphone are

People are likely to wait in long lines for their brainphones.

right now.

The idea sounds *crazy*, as in "over the top," in that it is such a milestone for humanity, albeit a bad one. Humankind's milestones over the last five million years include harnessing fire for good use, inventing and mass producing the automobile, creating a vaccine for polio, inventing the telephone, traveling in space, and making use of the home computer. Having processing and communicating technology inserted in the human brain for relatively permanent use goes along with this list to the degree that it advances our capabilities and it sociologically alters us. But it runs contrary to the list in that none of the other items physiologically alters us. Even vaccinations, which admittedly cannot be reversed, only produce antibodies in our blood, but they do not change us physically, mentally, or behaviorally. The interweaving of human tissue and machine technology is a major, major anthropological event. It *is* crazy, but it's not so crazy that it's not going to happen—unless we choose to make some big choices over the next few years to keep it from being so.

When addressing why the brainphone is in our near future and how it could forever change the human race, it merits looking at our innate yearning for technology and connectivity. The human desire for advancement is built directly into our DNA. MIT research scientist Lex Fridman, recently asked by comedian/podcaster Joe Rogan if the insatiable human thirst for technology resulted in the nuclear bomb, replies, "I don't think it's possible [for humans] to *not* build a nuclear bomb…There's something about us chimps in a large collective, where we are born and pushed forward towards progress of technology. You cannot stop the progress of technology. So, the goal is to…develop how to guide that development into a positive direction" ("Joe Rogan Experience #1188," 2018). Encyclopedia author Charles van Doren, in *A History of Knowledge* (1991), makes a similar observation:

> Many human beings are unhappy either because of what they know or because of what they do not know. Ignorance remains bliss only so long as it is ignorance; as soon as one learns one is ignorant, one begins to want not to be so.…The desire to know, when you realize you do not know, is universal and probably irresistible...It is impossible to slate the thirst for knowledge. And the more intelligent you are, the more this is so.

Perhaps the best commentator of all in this regard is Hollywood writer/director/producer James Cameron, who has Arnold Schwarzenegger's T-800 robot make this in-depth analysis about the human condition in the 1991 film *Terminator 2: Judgment Day*: "It's in your nature to destroy yourselves."

Yeah.

Of course, it is equally human to view the present as much worse than the past and to yearn for a time that probably never was. And so, it is worth asking if, perhaps, the craving for new technology and better interconnectivity is no worse than, say, desiring the automobile back in the early 1900s. One possible answer is that there is a distinct point in one's experiences when a craving becomes an addiction. And the addictive nature of smartphones and social media has been well documented.

We Will Stay Addicted to Smartphones

Do we simply crave our smartphones and their many features and applications, or have many (if not most) of us entered a period of addiction? The distinction is important: one only has to look at the *before* and *after* photos of people addicted to methamphetamine to see what humans are willing to put themselves through to avoid the mental and physical pains of withdrawal. Even among young users, meth addicts experience dramatically high rates of poor health, heart disease, hypertension, asthma, ulcers, and death (Herbeck, Brecht, & Lovinger, 2014). Yet many of them continue on for years and years taking meth and suffering through these symptoms as they pick at their faces and watch their teeth fall out. Will the person with a piece of iGen machinery inserted in his or her skull be any less damaged than, say, the meth addict in that Year Four photo? Current research suggests the answer is likely "no." Here's why.

The definition of *addiction* seems, in some circles, to be getting murkier over time, but it really shouldn't be. The American Psychiatric Association (APA), while acknowledging some of addiction's complexities, offers a very straightforward meaning, calling addiction "a brain disease that is manifested by compulsive substance use despite harmful consequences" (APA, 2017). That is, a person's mind gets sick, and he or she does something over and over again past the point where it becomes damaging. Changes within the human, addicted mind are neither nuanced nor nanoscopic: they are physical and obvious. Writes the APA: "Brain imaging studies show changes in the areas of the brain that relate to judgment, decision making, learning, memory and behavior control" (2017).

The only thing slightly controversial (but certainly supportive of the arguments in this book) is APA's recent acknowledgment of brain-produced dopamine as a potentially addictive natural substance. That's why the fifth edition of the APA's *Diagnosis and Statistical Manual of Mental Health Disorders* (*DSM-5*) covers new categories of addiction to include compulsive gambling, Internet gaming disorder, and caffeine use disorder.

APA's inclusion of the word *disease* in its definition has been an occasional topic of conversation with my colleagues over the

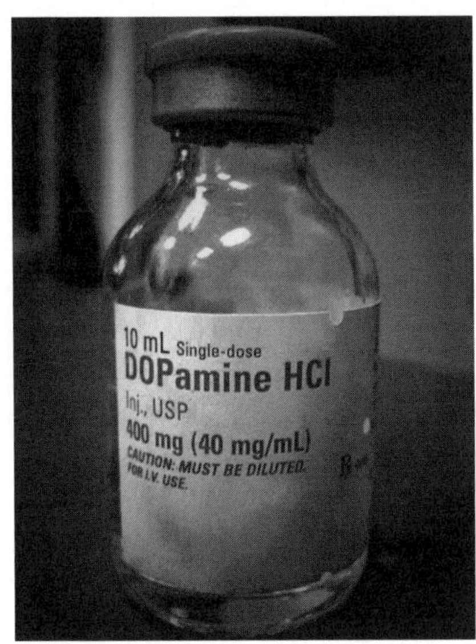

Dopamine, a chemical produced naturally in the brain, is now considered potentially addictive. Photograph by Intropin. Offered for use via Creative Commons. License https://creativecommons.org/licenses/by/3.0/deed.en

years. Some of them have had difficulty regarding addiction as identical to, say, a disease such as cancer. For people who have not experienced addiction themselves via their own life story or the story of a family member or friend, the connection might not seem apparent. On the other hand, for those of us who have watched both addiction *and* cancer sicken and eventually kill different family members, the similarity of progression is striking. Upon medical inspection, addiction does, indeed, show several attributes of a disease:

- Medical professionals can identify addiction through the traits of the patient. (Symptoms leading to a diagnosis.)
- Addiction tends to be inherited. In other words, an addict tends to have had a parent or a grandparent who also was an addict. (Family trait.)
- If the addiction is left untreated, medical professionals can pretty accurately predict what will happen. If the addiction is treated, medical professionals also can pretty accurately predict what will happen. (Prognosis.)

- There are concrete, evidence-proven ways—such as attending a Twelve-Step Program—to control addiction. (Treatment.)
- As with cancer, following treatment, the symptoms of addiction may return again down the road. (Possible relapse.)

Which brings us back to our initial question: Have we entered a period of addiction to our smartphones? Yes, yes, yes, say the researchers. Let's take the five-point template from above, which explains why addiction is like most other diseases, and juxtapose it against what researchers have discovered about young people using their smartphones. The results are surprising and terrifying:

- Average smartphone users touch their smartphones 2,617 times per day. The top 10 percent of users touch their phones an average of 5,427 times per day (Winnick, 2016). They are dramatically less likely to leave their bedrooms and go outdoors. They are dramatically more likely to feel lonely. They are more likely to feel useless and to view themselves as failures (Dr. Jean Twenge research, 2017). In Australia, sports scientists are monitoring small bone spurs, resembling horns, growing out of the backs of some young people's skulls, and are considering the possibility that the spurs are caused by young people staring down at their smartphones all the time (Shahar & Sayers, 2018). (Symptoms leading to a diagnosis.)
- On average, 26 percent of daily smartphone usage is spent using Facebook and sending text messages to people we know (Winnick, 2016). (Family trait.)
- Left unchecked, a young person who remains addicted to his or her phone is, on average, 50 percent more likely to be clinically depressed than a similar-aged person from a previous generation without smartphones. A smartphone addict in his or her young teens is twice as likely to commit suicide (Twenge, 2017). (Prognosis.)
- Limiting smartphone use to under two hours per day dramatically reduces the negative physical and mental impacts of the addiction. Exercising, going outside once a day, spending time talking face-to-face with people (in

person—not FaceTiming), and getting a good night's sleep all help reduce the compulsive urges an addict has to be on one's smartphone (Twenge, 2017). (Treatment.)

- Medical professionals in South Korea are attempting to treat smartphone addiction in young people in clinics similar to drug clinics. Researchers have found that therapeutic recreation, exercising, music therapy using drumming activity, and art therapy are helping young people to keep from returning to compulsive use of their smartphones (Kim, 2013). (Possible relapse.)

Any way you slice it, at least one generation of smartphone users is becoming addicted to their devices, and this addiction, like a disease, is eating away at their lives and their bodies. In many cases, the disease is killing them.

We Will Stay Addicted to Social Media

Since most people access social media, such as Facebook, through their smartphones, it is easy to equate addiction to

Teens addicted to their smartphones are 50 percent more likely to be clinically depressed than those from a previous generation. They are twice as likely to commit suicide. Photograph by Tammy Gann. Courtesy of the photographer via Unsplash.

smartphones with addiction to social media. However, there is a slight, but noteworthy, difference. Addiction to social media is distinct in that: 1) It includes the extra factor of constantly comparing oneself to others; 2) Its real-time elements include the potential for staying up throughout the night in order to stay in a social loop; and 3) It includes the potential for being bullied.

Soft skills trainer Bailey Parnell (2017) tells an interesting story about being on vacation in Jasper National Park of Canada and trying to go without posting photos along the way on her social media. In her lecture, she discusses finding herself feeling anxiety and withdrawal symptoms from not accessing her social media. She says those feelings didn't begin subsiding for nearly four days. And those feelings are from being *off* social media. They don't include the common stressors, Bailey notes, that come from being *on* social media, to include unfairly comparing one's real life with the highlight-reel nature of other people's postings; worrying about not having enough social currency through likes, comments, and shares; fear of missing out; and being harassed online by others.

Again, let's take the addiction-as-disease template and hold it up against what researchers say about people—particularly young people—and their social media:

- Teenagers who spend these large amounts of time on social media are struggling. They are more likely to not sleep enough. They are twice as likely to feel generally unhappy as the generation before them (Twenge, 2017). (Symptoms leading to a diagnosis.)
- On average, Facebook accounts for 15 percent of all the time we spend touching our smartphones (Winnick, 2016). (Family trait.)
- Young people who do not curb their social-media usage are nearly twice as likely to exhibit at least one risk factor for suicide (including feeling sad or hopeless for more than two weeks) as the generation before them (Twenge, 2017). They are more likely to become obese, to see their school grades drop, and to engage in risky behavior (American Academy of Pediatrics, 2016). (Prognosis.)

- The American Academy of Pediatrics (2016) offers ways to prevent or control social-media addiction, to include being forced to shut one's phone off in the evening; planning media-free time and places in one's home; playing sports; and regularly engaging in face-to-face conversation. (Treatment.)
- Between 4 percent and 8 percent of children and teens who try to control their social-media use aren't able to. The number might approach 10 percent for online gamers (2016). (Possible relapse.)

Admittedly, a person can spend too much time in front of *any* type of screen—television, DVD player, video game screen, computer, digital notebook, or smartphone. The differentiation with social media is the interconnectivity, not just with other people but with collectives of people. American psychology pioneer Abraham Maslow argued that, after food, shelter, procreation, and safety, there was no human need more basic than the need to belong and be accepted. Feeling accepted means getting positive feedback. When a person posts something on his

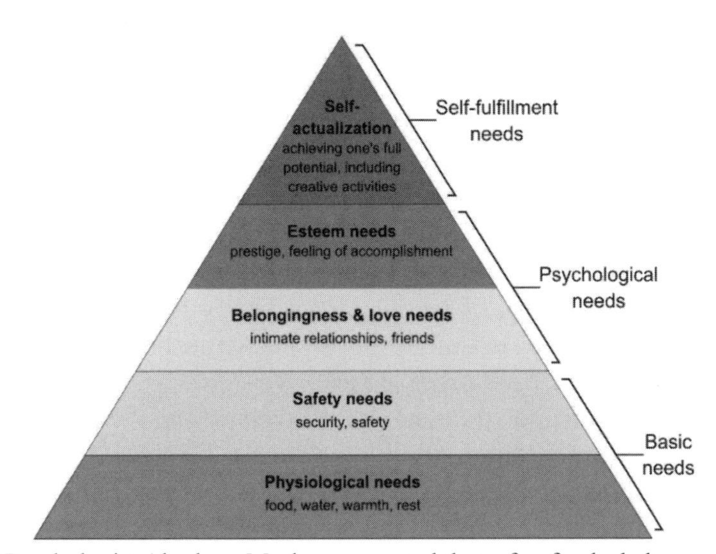

Psychologist Abraham Maslow suggested that, after food, shelter, procreation, and safety, belongingness is the most basic of human needs. Chart by Android Mars Express. Offered under fair use via Creative Commons. License https://creativecommons.org/licenses/by-sa/4.0/deed.en

or her social media account and receives a "like" or any other type of positive acknowledgment, the dopamine rush is wonderful—and potentially addictive. Apps are designed that way, to keep the dopamine dripping as if from an IV bag, and to keep us all hooked.

In many of the Twelve-Step Programs that exist to fight addiction, there's another, simpler definition of addiction. It's in the form of a question: Is whatever you're currently doing too much of (drinking, drugs, smoking, gambling, sex, gaming, etc.) now making your life unmanageable? That's certainly a good question for the person who spends hours a day on a social-media account to ask himself or herself: "Is spending so much time on these media apps affecting my family life, my work life—even my health? Have I gotten to the point where I'm now defining my successes and happiness by how I appear to others on social media? Is worrying so much about how I look on social media affecting my *real* life—how things *really* are and whether or not I'm *really* successful or happy?" Some corporations, of course, would love for you to continue to define your life through how you appear on your social-media presence. In order to do so, you have to be on *their* media, watching *their* sponsors, learning from *their* so-called news items, and striving to be part of *their* in-group. In other words, existing in *their* reality. As with the addicts who fight against corporate alcohol, corporate tobacco, and corporate narcotics, the addicts who gather nightly to fight against social media understand the power of the corporate entities that want to keep them hooked. Understanding the enormous influence of big business, these addicts seek a Higher Power (whatever that might be for them) to help them in their quest to call their own shots in life.

One more quick thought about Twelve-Step Programs. Many addicts, when sharing with the group in a Twelve-Step meeting, often will mention how they had to *hit rock bottom* before coming around to realizing that they needed help. That is, their lives had to sink to the rock bottom of the well—lost jobs, lost houses, lost relationships, near-death experiences—before it finally hit them that they were sick and needed to heal. In fact, when dealing with a family member who has become an addict, people often are warned that their loved one may, indeed, have to hit rock bottom before they'll come around. People sometimes are advised not to lecture or enable the addict, but, instead, to allow the addict to hit rock bottom with the hope that the person might have an epiphany moment and begin to turn his or her life around.

One of the many problems with the brainphone is that there can be no epiphany moment to step back and turn oneself away from what's causing the addiction, because what's causing the addiction is attached! The scenario is similar to someone hooked up to an automated insulin pump that dispenses liquid nicotine instead. How could anyone ever give up nicotine addiction while connected to such a device? And what corporation would want them to?

We Will Continue to Accept Intrusion

If a stranger knocks on your door tonight and asks to come in and discuss your life with you, will you let him in? What if he wants to go up to your child's bedroom and have a similar discussion with your child, without you tagging along? Will you allow him? No way. And yet, in many ways, we do these things every night. We allow people with only minor (or false) connections to our friends into our social-media circles. And we allow our children to converse regularly, via social media, with complete strangers.

The American Academy of Pediatrics (AAP) (2016), in listing the eight main problems they have with people spending too much time on social media, suggests that half of them likely relate to one's communication with strangers. These problems include learning risky behavior—such as self-harming or purging one's food—from strangers on social media; sending or receiving sexting; being sought out by predators; and being cyberbullied.

It's not easy to address these problems. Essentially, parents are trying to counter a situation at a time when their young-adult children think they know everything. (Heck, we all thought we knew everything at that age.) They're belligerent, and they're extremely secretive. Plus, since they're growing up with the technology that many of us are still learning, they're generally three steps ahead of us as we try to monitor them and screen the different apps that they're availing themselves to. Talking to them calmly, as the AAP suggests, about online citizenship,

Alarmingly, about half of all problems related to children communicating on social media involves them communicating with a stranger.

respect, safety, and privacy is tough when they're throwing up their arms and storming out of the room in disgust.

It's also tough to discuss with our children their careless online behavior when we, as adults, do such a poor job and set such a poor example. Many of us spend just as much time on Facebook, WhatsApp, Instagram, Snapchat, and Twitter as they do (although we lean towards LinkedIn and they lean towards whatever's new). And we may be slightly more vigilant in keeping total strangers out of our social-media circles, but we tend to tell people in our circles, including people we only have passing relationships with, just about everything about ourselves. Many of us have extremely detailed resumes or curricula vitae posted online. We tell people when we're on vacation (which means we're letting everyone know when our houses are empty). When we send photos and/or locations, they include the exact map coordinates of where we're standing. We complain about our bosses (figuring, I suppose, that it will never get back to them). Many of us post photos or videos of ourselves drinking and making idiots out of ourselves. Some of us post photos or videos or narratives of ourselves committing crimes or admitting to committing crimes.

Any one of these things is forgivable and generally reversible. Unless people begin frantically reposting something you would rather not have them see, you always can remove the item from your social media as if you never made the mistake of posting it in the first place. The problem with today's intrusion is that corporations and governments are not simply interested in that one item: they are interested in patterns of items. So, when you post the unfortunate photo, the action is recorded. And, when you remove it, *that* action is recorded. And, thousands and thousands of touches, clicks, or swipes later, an organization can have quite a good profile on you—for selling things to you, comparing you with other people, convincing you to behave in a certain way, or for just keeping tabs on you. If they are a corporation, they might offer your profile to another corporation for money. If they are a government, they might offer your profile to another government. Because, after all, we never can have too many Big Brothers watching us.

It's interesting what social-media companies keep track of these days. In order to join a particular online community, we all need to provide our names, genders, contact information, and dates of birth. That information, right there, is pretty personal, when you think about it. But the intrusion digs much deeper. As you innocently post pictures and videos of yourself and offer positive responses to your friends, the company is tracking what types of photos and videos you're generally sharing and what, on average, generates positive responses from you. The company keeps tabs on what advertisers you check out. They watch where you go from their site, via links. They track where you've lived and worked. They record what groups and organizations and other social-media circles you belong to. They check out the other people in your circles of acquaintances, what *their* backgrounds are, and how their backgrounds correspond to *your* background. If you use your social-media membership ID to access another site, the company that runs that other site is often allowed access to all of these tidbits of information as well.

The terms these companies use to gather all of this information on you and form a profile on you is a little chilling. The social-media companies call it *mining*. Other companies that extract the data—sometimes legally, sometimes illegally—call it *harvesting*. Feel like cattle? You should.

Just as chilling—and yet, just as generally accepted—is how easy it is for a corporation, or even another individual, to track our Web browsing histories. No matter how strong our privacy settings are or how often we clear the browsing histories in our phones or on our laptops, it is relatively straightforward to identify a person and figure out which websites he or she has visited over time.

In 2017, German journalist Svea Eckert teamed up with German data scientist Andreas Dewes to see how easy it would be to acquire the so-called "anonymous" Web browsings of a few million people and then figure out who some of these people were. ("'Anonymous' Browsing Data," 2017). They pretended to be a marketing company trying out a new algorithm for determining what buzzwords were working online. A data broker fell for the ruse and offered them the browsing histories, supposedly de-identified, of three million Web browsers in Germany.

Then Eckert and Dewes went about the task of identifying the de-identified people. It was easy, really. If a company looks at the websites you visit today (your employer, your bank, your library—perhaps even the search results for your own name on Google!), it is probable that they will figure out who you are. Even without your name included among the browsings, Dewes suggests that, mathematically, it only takes visits to 10 different websites to determine who the person is.

As it turned out, Eckert and Dewes were able to identify 30 very prominent people and what places on the Web they each had visited. They were able to tell which porn sites a judge regularly visited. They were able to figure out what medicines a German member of parliament was taking. And so on. Fortunately for these 30 people, this journalist and data analyst had good hearts, and so they privately informed the people what they had found, but they did not make their names or sites-visited public.

Eckert and Dewes presented the results of their investigation at the Def Con hacking conference in Las Vegas in 2017. And the world collectively yawned. I have to wonder how many people in the audience missed the presentation entirely because of whatever they were looking at or doing on their smartphones.

The truth is that, when it comes to smartphone and social-media intrusion into our lives, we're all naive, we're all childlike. We accept it. We become numb to it. Perhaps the biggest outrage of all is that there's no outrage.

We Will Continue to Fear the Callout Culture

Some of us are old enough to remember the old chat rooms of America Online (AOL). Hearing that burst of data on the telephone's landline, wired over to the computer. Watching AOL come up on the screen. Checking one's email. Instant Messaging (IM'ing) a friend. And then checking out the chatrooms! How cool! I remember entering a chat room regarding politics. My initial thought when all this started was, "What a wonderful venue for exchanging different ideas, expressing one's views freely, and enjoying intelligent, political discussion!" Don't laugh. As you might guess, that thought didn't last long. What I generally saw on the screen was one opinion gaining traction quickly, without many facts to back it up. Then, anyone who questioned that now-dominating opinion was berated and shouted down. (ALL CAPS!!!) Many times, the person offering the opposing view was labeled ignorant. That is, he or she was "called out" for being an idiot. And thus, the online, *callout culture* was born. Of course, some people were in those chat rooms just to stir up trouble. (We now call them *trolls*. As an aside: Is it callout culture to call a troll a *troll*? Just wondering.) But the groupthink belonged to nearly everyone.

Incidentally, something else happened in those chat rooms. Since, initially, people utilized the front part of their AOL email addresses as their names in the chat rooms, anyone who wanted to could collect the email addresses of dozens of people with similar interests, and, later on, send them a mass email trying to inform them or to sell them something. And thus, spam was born.

Callout culture and the mass hysteria that goes with it are nothing new. For example, the Communist overthrow of nations generally begins with convincing the masses that anyone trying to be successful by earning a dollar is a demon, then whipping the masses into a frenzy, and then having the masses attack the demons. The converse also occurs—convincing the masses that anyone who questions American capitalism is a demon. In the 1950s, Wisconsin U.S. Senator Joseph McCarthy conducted his Red Scare hearings and ruined many people's lives. To a smaller degree, labeling activists and left-leaning politicians as Socialists continues today.

The trouble with any label is that it places a person into a category, packages that person in an over-simplified manner, and then places a bright bow on the package for anyone to see clearly and disdain from a distance. Surely every single person on Earth is more nuanced than this. Most of us have opinions that are all over the political map—with many of these opinions being situation-specific. Many of us have changing opinions that become better informed and transformed over time. So why the label? Why the hostility?

As it turns out, the hostility is the point. Many passionate people these days believe that labeled people, such as racists, misogynists, and nationalists, can't be reasoned with and shouldn't be approached with debate or discussion. They should be "called out" until they "wake up" (or "are woke"). The notion is that the public, social-media shaming will force them to reckon, on a personal level, with who they are, and that they will change for the better. Anyone with any sort of training in psychology knows that this strategy is folly. Sure, the loud, public shaming might force the person to outwardly change, for fear of sustaining further criticism or enduring more social (and social-media) rejection. But inwardly, the person is likely to harbor resentment and hold onto his or her convictions even more strongly—and when people are quiet and careful about what they believe, they can do much more harm. It's also worth mentioning that psychologists suggest the experience of being shamed is, in itself, a traumatic experience. So, in aggressively calling people out, we might be turning outcasts into, well, outcasts with emotional problems.

Another reason for the over-the-top hostility: People raised in their cyber-worlds are having a difficult time differentiating virtual reality and real-world reality, and they are finding it very hard to handle their feelings. As a result, they are beginning to view opposing opinions as physical threats to themselves, particularly at universities and in their workplaces. Attorney Greg Lukianoff and psychology professor Jonathan Haidt (2018) lay out, in detailed, well-argued terms, this thesis in their book, *The Coddling of the American Mind.* They explain how staying on smartphones and on social media all the time makes people hypersensitive and hostile to counter-intuitive viewpoints. Furthermore, people's existence in an online, neo-actuality encourages them to follow their feelings and assume that a threat exists whenever they have feelings of discomfort—essentially discounting anything society has learned over the years about mindfulness and established, cognitive-behavioral theory. Universities and companies would be well served by forcing these extremely touchy people to engage their opponents in thoughtful discourse. By doing so, they also would be well serving the American values of freedom of thought and freedom of speech. Instead, however, these organizations—worried about lawsuits, their own branding, and being called out themselves on social media—cater to these fragile souls and do their best to insulate them from most anything they find upsetting. Coddling them, if you will. Write Lukianoff and Haidt:

> Sages in many societies have converged on the insight that feelings are always compelling, but not always reliable. Often, they distort reality, deprive us of insight, and needlessly damage our relationships. Happiness, maturity, and even enlightenment require rejecting the Untruth of Emotional Reasoning and learning instead to *question* our feelings. The feelings themselves are real, and sometimes they alert us to truths that our conscious mind has not noticed, but sometimes they lead us astray. (2018)

So, then, people get lost in their cyber-world, become hypersensitive to anything that runs contrary to what they deem right in that world, and then get hostile both in their cyber-world *and* in the real world to people whose opinions or statements make them uncomfortable, as if the words themselves were physically injurious. If you haven't recently witnessed these types of people and behaviors as described, then you haven't been on a college campus or in a large workplace in a while.

People living in cyber-worlds are having tough times handling their feelings. Photograph by Christian Erfurt. Courtesy of the photographer via Unsplash.

The trouble with smartphones and social media as they relate to callout culture is that people nowadays are always *on*, always connected, and, collectively, always ready—and eager—to pounce. The feedback and instant outrage and hostility is loud and threatening. Also, since many children and young adults—wrapped up in their social media and gaming—are still living with their parents and slow to mature, the feedback, frankly, is often silly and childish. But it's still a threat to people's livelihoods and their right to hold opinions. And so, most of us either embrace the opinion that's least likely to produce the combative, collective response, or we quietly toe the line and hope no one notices us. The good news at present is that it is still possible to shut off our devices and our social media (not easy, as we're addicted, but still possible) and not have to worry much about the electronic calling-out bleeding into our physical realities. The bad news is that, once the brainphones are in, and our thoughts are intertwined with cyber-worlds, it will be much more difficult to step away from our electronic existences.

CHAPTER 3:
HOW WILL WE GIVE IN TO THE ADDICTIONS OF CORPORATIONS AND GOVERNMENTS?

No one can serve two masters; for either he will hate the one and love the other, or else he will be loyal to the one and despise the other. You cannot serve God and mammon.
(Matthew 6: 24 NKJV)

We Will View Corporations as Our Friends

There are two parts to the upcoming brainphone storm—addiction to technology and addiction to power. The second part of the storm, addiction to power, stems from how corporations and some governments are hooked on control. They want to have it all, and they will never have enough. Intentionally or not, their systems of growth have worked systematically to destroy the middle class and to oppress vast numbers of people throughout the world. As with all addictions, the thirst for power is never quenched, but pushing brainphones will help them feel like they're getting close—watching and controlling, via technology, what everyone sees, thinks, purchases, and believes.

I have watched the results of addiction to power up close in three regards. First, I have seen people struggle and lose the fight to stay in what most people would consider to be the middle class. They borrowed heavily, and then they lost the house they had borrowed against. They took on second jobs (in some cases, second full-time jobs, working 80 hours a week). They lost solid manufacturing jobs, as those factory operations were moved to countries where exploitation of human labor (horrible pay, long hours, unsafe or poisonous work conditions) was allowed or ignored. Whole neighborhoods that used to be considered deprived were now filled with struggling former middle-classers, in many cases working very hard just to stay in their homes. So, if the middle-classers were moving into the poor neighborhoods, where were the impoverished going? Tents! I watched many tent cities go up in the United States. Eventually, people in nearby homes complained, and police came along and cleared everyone out. The tent people moved on, eventually finding another wooded area or abandoned parking lot in which to relocate.

Was it all by design? Certainly. Those middle-class jobs went away because companies collectively found parts of the world where they could take advantage of people. If they were lucky, they could set up shop in a country that essentially let them enslave people. The rich stockholders calling the shots? Oh, they stuck around here, moving into nicer houses. Was it a conspiracy? Not really. By most definitions, a conspiracy is unlawful and planned in secret. Powerful people getting rid of the middle class to become more wealthy and powerful did so legally in the United States, in full view of everyone. More on this phenomenon, and why it matters, in a bit.

The second thing I've watched, up close, suggesting an addiction to power, is a consolidation of mass media. As a writer and a former young radio broadcaster, I'm concerned by this trend. For example, I watched a successful publisher of mine get swallowed up by a much larger publishing company. That publisher, in turn, was purchased by another, which was purchased by another, which was purchased by another! The result is a giant multinational conglomerate publishing company, Penguin Random House. So, it stops there, right? Maybe not. There are currently five major publishing conglomerates in the world. This number might be reduced down to four (for now), with the massive mega-purchase of Simon & Schuster by Penguin Random House. At this writing, the merger is being reviewed by the United States federal government for possible anti-trust issues.

Print media is not the only mass media area of mergers and consolidation of control. Television, entertainment, Internet media, and social media all have experienced similar trends. "But Scott," you might ask, "what about all those self-starters with their own podcasts or their own highly successful social-media pages? Doesn't the diversity of all those commentators and influencers run contrary to your argument of the 'powerful few?'" Uh, no. While I do admit that several hundred television channels are better than three, the truth is one very large media corporation likely owns and controls many of those channels. As for the independent podcasters: They are not sending out radio waves from an antenna on their roof; they are using one of a relatively few online platforms to get out their message. And they must follow the rules of the platform in order to stay on the Web and to stay relevant. Sometimes, they are removed from those venues for very arbitrary and unfair reasons.

The third thing I've watched, up close, is a merging of technology companies. A few years ago, I was hired by a small cybersecurity company with about 250 employees. Shortly after I began to work there, our company was purchased by a large publicly held corporation of about 14,000 workers. The company sold, amazingly enough, for $250 million. That's $1 million per person—who knew that I was so valuable? Anyway, a little over a year later, that large corporation was purchased by a private equity company, managing tens of billions of dollars in assets, for $7.1 billion dollars. What a culture shock! My head is still spinning, and I cannot imagine what is coming next.

That is a *lot* of money. Sadly, it is money not going into new market expansion, new ideas, new sectors, or new jobs. It is capital simply being used for acquisition, for control. And yet, most people tend to treat such mergers in friendly fashion. A company that expands through acquisition is, on paper, the same as one that expands by gaining customers on its own. The result on paper is one of growth, reflected favorably in stock prices and everyone's 401(k)s. Mergers suggest efficiency by weeding out duplicate jobs. Plus, since larger companies can purchase supplies in greater bulk, the economies of scales look good to stockholders as well. In fact, a large enough company can purchase its suppliers or distributors, in a vertical merger that also impresses stockholders.

But the truth is that many mergers happen so that large companies might eliminate worthy competition. That is, they remove their rivals by buying them. Many times, a large company recognizes that a small company is attracting all the new, innovative talent and coming up with all the new, cool ideas and products. Rather than addressing and trying to change its stale company culture into something contemporary and pioneering, the large company simply purchases the small

Many large companies merge to remove the competition. Photograph by Walter del Mundo. Courtesy of the photographer via Unsplash.

company in order to acquire the young talent and their ideas. Does this bring a new vibrance into the big company? No. Once the merger is finished, the innovators and smart thinkers, not relishing the idea of belonging to an old, stale behemoth, run off to join another small, flexible company, or they form their own, new, small companies. Those who stick around are assimilated into the large company's existence and culture ("Resistance is futile."), never to be innovative or appreciated again.

I certainly don't begrudge corporations for trying to grow, and I don't begrudge stockholders for trying to make a buck. But the breakneck speed of these mergers, not to mention the exponential nature of the growth, suggests to me an addiction to power. Again, *addiction* easily is defined as doing something excessively to the point where life becomes unmanageable. While mergers aren't unmanageable, they make a lot of people's lives so. New email systems, new payroll systems, new chains of command. There are layoffs, buyouts, combining of departments, combining of individual job responsibilities. And a year later, it happens all over again. And to what end? At some point, managers begin to fail as their spans of control become too large or too poorly defined. The company becomes less (not more) efficient as diseconomies of scale set in. And everyone becomes too darn burnt out. And, like the cherry placed on top of a manure sundae, at some point the organization is investigated for possibly becoming a monopoly. All of these eventualities are predictable. So why do corporate boards seek mergers and more and more power if the end result is an unmanageable state? Hmmm, what did we say the definition of *addiction* is?

Incidentally, it isn't just the corporation-as-organism that's addicted to wealth. Studies have suggested that rich people as individuals are equally prone to the disease of wealth addiction. Harvard Business School professor Michael Norton interviewed two thousand millionaire/billionaires in 2018 and discovered that their added wealth was making them less and less happy with each additional million, much as someone addicted to alcohol progressively needs more and more drinks to get the same buzz (Donnelly, Zheng, Haisley, & Norton, 2018). That is to say, money, like most everything else, has a diminishing return on the happiness it brings you, real and imagined. And too much makes your life unmanageable. "All the way up the income-wealth spectrum," Norton told *The Atlantic* about his study, "basically everyone says [they'd need] two or three times as much" money to be perfectly happy (Pinsker, 2018). I remember talking some years ago to a wealthy, multi-millionaire friend of mine who admitted something very similar: "Scott, I don't think I'll ever feel secure until I hit a billion." If he ever gets there, I hope he lets me know if it worked! (I'm doubtful.)

But contrary to the smelly town drunk—clearly addicted to alcohol—who people cross the street to avoid, we rarely avoid the tycoon or corporation. Much to the contrary, many of us are drawn to corporations rather than being wary of them as entities addicted to wealth and power. We obliviously look at them as friends of a sort. We wear their logos. We *like* their posts on social media. We revere their super-rich (and equally addicted) leaders. As stockholders, we appreciate their practices, such as making aggressive acquisitions, as strategies for increasing our shareholder values. We let them take care of us. Finally, and perhaps most importantly, we let them tell us what the trends are and what to purchase.

We Will View Governments as Our Friends

You may or may not be old enough to remember President Ronald Reagan. If you remember him, you may or may not have liked him or his policies. However, in all likelihood—like or dislike—you probably would agree that he was very good at messaging. He ran his Administration on three or four main messages (Communism is bad; too much regulation is bad; etc.), and allowed others to work out the details. His public-speaking skills were powerful, resulting from years in professional acting and years running a union and, eventually, the state of California.

Reagan's main message as presidential candidate and as president was, essentially, that government was bad and had to be reduced in its size. That is, if people were freed from the constraints of too much taxing and too much regulation, their lives would flourish. Many people loved the notion of a life less regulated, and Reagan was permitted by both Republicans and Democrats in Washington, DC, to cut taxes and whittle away at regulations and restrictive government offices.

After eight years of a Reagan presidency, what was the result? The size of the United States federal government was larger than ever. It had grown in the number of federal employees and as a percentage of gross domestic product (Labonte, 2010). And, although some deregulation had taken hold, other attempts tended to have a long-term backfiring effect, such as the deregulation that resulted in many savings and loan companies collapsing and the Federal Savings and Loan Insurance Corporation becoming insolvent. Some refunding and re-regulation had begun before Reagan's presidency concluded. While some deregulation took hold long-term, such as in the airline industry (to worthy, economy-pricing effect), Americans still ended up largely at the hands of their federal government. I personally would argue that such outsized control of the government exists in America today.

The point of bringing up Reagan is to suggest that reeling in the domination of big government (along with the domination of big corporations) historically has not been an easy task. If someone so inspiring and charismatic and, indeed, powerful in his own right, couldn't reduce the size or control of an ever-expanding federal government, I'm not sure anyone can.

So why do the federal governments of most advanced countries on Earth continue to grow unchecked? For communist countries, religious states, and dictatorships, the question answers itself. The larger the government, the easier it is to control the people. On the other hand, for democratic countries, the question is not easily filleted. If the voters want a less-oppressive government, can't they simply vote out the politicians who support greater bureaucracy? The answer is "they can," but the quandary is that voters, who get what they want, often don't really know what they want. When this country was attacked on September 11, 2001, by al Qaeda terrorists, I don't recall anyone discussing reducing the size of the government. The same for the economic collapse (related to the real estate crash) of 2008. The same for the COVID-19 pandemic of the 2020s (which continues at this writing). To the contrary, most people I spoke with were wondering what the federal government was doing about the situation. And, in each case, do stuff they did, to the tune of trillions of dollars for each calamity.

The problem with voters (okay, perhaps *problem* is a strong word) is that, once the government responds to a disaster by putting remedies in place, people tend to like the remedies and want to keep them around. Political economist Robert Higgs (1990) makes the argument that, once the United States government dramatically grew in size and scope following the Great Depression of the 1930s and then World War II, it maintained its enormous reach and, in fact, continued to grow unchecked. Many groups of people (including rich people and rich corporations, by the way) benefitted from the economic giveaways tied to this control, and they became politically active to see that these gifts were kept in place. Writes Higgs:

> [Following World War II, in the United States] there remained no fundamental check on the growth of government—nothing to perform the restraining functions that the old Constitution and the dominant, limited-government ideology had performed in the 19th century.

> Politicians now could offer to sell virtually any economic policy whatever, no matter how few the gainers and how many the losers. Of course, such conditions were tailor-made to bring forth special interests prepared to buy the policies from the politician-suppliers.

In other words, as long as all of us are getting something out of the government, and as long as some of us are getting lots out of the government, we tend to let the government continue to grow and take over. As of the most recent, available U.S. Census statistics, federal, state, and local government workers make up one out of every five workers in the United States, not including full-time military personnel (U.S. Census Bureau, 2010). We tend to accept this unrestrained expansion willingly and blindly: government becomes an intricate, seemingly beneficial part of our lives. Will we allow the trend to continue when the government finds its way into very personal and physical aspects of our lives? Will we lie down and stay still when the government tells us it's time to get our brainphones?

One in five U.S. workers is employed by a state or local government or the federal government, not including the military. Photograph by Clark Van Der Beken. Courtesy of the photographer via Unsplash.

I used to think that there was a point where Americans might push back against an ever-expanding, ever-reaching government. After all, we can't *all* become government workers, can we? The percentage of public sector vs. private sector can't *really* reach 100/0, can it? Nowadays, I'm not so sure. If we consider the federal government using the corporation-as-organism view, we find that it certainly has the same, collective mannerisms suggesting an addiction to power. And, within the government, politicians and administrators also display these traits. Furthermore, they know what gets them power and keeps them in power: funding. Presumably, the more funding, the more power.

It's also worth mentioning that the government is filled with people who, by virtue of their formal job descriptions, know how to manipulate the masses. The White House alone, for example, employs hundreds of people in its press, communications, political strategy, and digital strategy offices. Furthermore, the military employs thousands of people in psychological operations (PSYOPS), whose sole purpose is to persuade the emotions, attitudes, motives, and morale of foreign populations that they're trying to control. Although they technically are not allowed to use these messaging tactics on the United States population, the line between what is acceptable and unacceptable surely is becoming blurred, as PSYOPS have begun to effectively intertwine with information technology. How is a PSYOPS specialist supposed to use covert or ambiguous psychological activity online in a way that guarantees it won't be read by U.S. citizens, or that U.S. citizens won't be impacted by the messaging? Young people entering the military are heavily recruited for the PSYOPS specialist role—they must be especially smart, and their training and experience are extensive. No wonder that they often are persuaded to leave being in uniform and join corporations, where they can bring a myriad of psychological and research skills to marketing and public relations departments (and make considerably more income.)

And so, between what the government may offer us in the way of money and services, and what they are able subliminally to convince us of through the talented skills of professional persuaders, it is no wonder that we have been lulled into the notion that government is our helpful friend. Yes, Big Brother is watching us, but he also loves us and wants to take very, very good care of us.

We Will Forget Human History

Are people sheeple by chance or sociology, or is it built into their DNA? There is a pretty strong argument that our tendency to follow orders—even the orders of those who want us to hurt one another—is a part of our natural wiring. History certainly suggests that humans have a funny way—Scratch that. A horrifying way—of doing what we're told. By assuming that the people who want us to get the brainphone have our best interests at heart, we will forget that history.

The future of the brainphone holds many questions regarding sinister behavior of one human towards another. Will governments lie to their citizens in order to convince them to receive the implants? Will people who refuse brainphones become objectified as subhumans? Will people who refuse them be forced to undergo the procedure? That is, will anti-brainphone citizens be stolen away in the middle of the night, anesthetized, and woken up the next day with working brainphones installed in their skulls? I suppose you know what my prediction is. Once one type of people objectifies another type of people, doing all kinds of bad things to those *subhumans* becomes easier, even justified.

As I mentioned in Chapter One, in relatively free countries, the social pressure and the employment pressure, alone, will be enough to prompt most consumers to purchase the implant. However, in oppressive countries such as China, Russia, and North Korea—where the brainphone will serve the State early on in very creative ways—citizens likely will be forced to have the procedure done. And it's not unthinkable that this coercion might become worldwide. If people are to be accepted into the collective, the procedure might happen at puberty or even at birth.

As humans, we've done some pretty horrible things to our fellow humans. Mass enslavement, mass imprisonment, torture, genocide, apartheid—the list is long. Notably, many of the crimes against humanity have been systematic and committed as part of government policy, with many, if not most, citizens participating willingly. "Not me," you think to yourself. "Those things happened long ago, and if any of them were to happen now, I wouldn't be a part of it." Really? Well, good for you. However, I respectfully suggest that you look at the articles of clothing you're wearing. Do you know what the odds are that one of those inexpensive but fashionable items was manufactured by slave labor in Asia? Very high. Not super cheap labor—slave labor. That is, migrant workers toiling without pay (a condition disguised at recruitment as "wages minus recruitment fees"), with their passports withheld so they cannot leave. At the very least, if we're purchasing these items without first doing some research on the brands and their contract history, we're complicit.

In wartime, human-on-human horrors get ramped up a hundredfold. During World War II, Nazi scientists performed hideous medical experiments on thousands of Jews and other inmates at the Auschwitz, Dachau, and Mauthausen concentration camps, including freezing, bone and muscle removal, poisoning, blinding, radiating, and the sewing together of living twins. Six million European Jews and millions of other inmates were mass murdered in camps throughout Germany. Just the task of moving all of these prisoners from where they had been living to where they would meet their nightmarish fates suggests a major logistical operation, involving thousands of willing civilians. It's tough for us to wrap our minds around the level of shock and suffering people brought down upon others, and why so many others went along. It's easier to forget history.

The script which convinces us that people who refuse the brainphone are somehow our enemies hasn't been written yet. But versions of it have been written and used successfully to wind up the masses throughout history, including relatively recent history. The script goes something like this: 1) Someone argues for something that is for the common good, but might come with some hardship; 2) The people most likely to lose out money-wise or power-wise plot against the person making the argument, if not the argument itself; 3) These people convince the masses that the initial argument is something meant to harm the common good rather than to help the common good; 4) The masses demonize and punish the person making the initial argument. The scenario devolves in a fashion similar to that of *An Enemy of the*

People, the 1882 play by Norwegian playwright Henrik Ibsen. In the drama, the protagonist, Dr. Thomas Stockman, ultimately is viewed as a villain, rather than as a savior, because the townspeople don't want to hear his bad news about the town's water supply and spas. Much to his surprise, he is demonized and ostracized.

The most recent example, at this writing, of such a phenomenon is the attempt by then-President Donald Trump to delegitimize the dangers—heath-related and economic—from the COVID-19 pandemic. His reasoning for doing so is still a mystery to me. (He wanted to turn the economy around before the 2020 election? He wanted to campaign without the hindrance of wearing a facemask? He wanted to celebrate the masculinity of those who chose to throw caution to the wind and live as if nothing was wrong?) In any event, he did what I viewed to be a remarkable job of ridiculing scientists, health officials, governors, and facemask-wearers in generals. At one point, there were certain pro-Trump taverns around my home where I wouldn't go for a coffee, because I thought entering or leaving with a facemask would start a fight—possibly verbal, possibly physical. (My fear waned considerably when Trump himself caught the virus and became deathly ill and hospitalized from it.)

The notion of shaming people or physically forcing people into getting brainphones is reminiscent of the *struggle sessions* during the Communist revolutions in Russia (during the late 1910s and early 1920s) and in China (during the late 1960s). A self-appointed tribunal (or just an angry mob) would show up at a person's home or place of work and would put the person through a public trial of sorts, subjecting the person to humiliation—even torture—until the person would admit to crimes against the community. The people targeted were often land owners or people doing well business-wise (ostensibly benefitting at the expense of their workers) or simply those opposing the political revolution. Many of them were killed; some of them were sent to concentration camps.

Psychologists have spent decades delving into why people are so easily convinced to do such horrible things to their fellow humans—even their friends and colleagues. In the 1960s, Yale social psychologist Stanley Milgram unnerved the academic

There will be "struggle sessions" for brainphone refusers.

community by showing just how easy such convincing is. In his experiments, people were asked to administer electrical shocks to others. (The shocks, in fact, were faked, and those being "shocked" were actors in on the sham.) With simple, guiding phrases, such as "Please continue," and "It is absolutely essential that you continue," people cranked up the amps and thought they had shocked someone to the point of passing out or dying.

Milgram was criticized for putting his subjects through the psychological discomfort of thinking they were zapping someone at high voltage—although the vast majority of them continued to do so. Milgram countered that the reason for the outrage wasn't his tactics, but that people were uncomfortable with the study's findings. Said Milgram (1974): "I'm convinced that much of the criticism, whether people know it or not, stems from the results of the experiment. If everyone had broken off at a slight shock or a moderate shock, this would be a very reassuring finding, and who would protest?" His discovery about people's response to figures of authority unsettled us into considering who we are and why, as humans, we are so easily swayed to hurt other humans.

In the case of brainphone indoctrination, those most likely to benefit from everyone having one are the tech rich (who want to sell the devices to the masses) and the government powerful (who want to monitor what we think). All they have to do is convince people that brainphones are somehow righteous and that anyone who refuses the implant is an Enemy of the People.

The Middle Class Will Vanish

I put forth the argument that any sort of awareness against the brainphone begins with the middle class. A healthy middle class contributes to a healthy, growing economy and makes democracy work. A society made up of only the very rich and the very poor is oppressive and, as noted above, potentially unstable. The people entrenched in corporations and governments do not necessarily have a problem with the "oppressive" part, due to their addiction to power, and they likely are not too worried about the "unstable" part. "Meet the new boss, same as the old boss," right? In one scenario, such future instability might ignite the desire for brainphones as a supposed remedy. *Get your brainphone and contribute to a new, stable society!*

If the middle class disappears, there is also the possibility that brainphones will be seen as a key to being one of the Elite. Initially the Party might offer the implant only to those who are controlling the politics or are running a community and are loyal to the Party. It might be seen as a status symbol. But that phase isn't likely to last: the desire by the government and corporations to have everyone plugged in, monitored, and controlled will be too strong.

In either scenario, a strong middle class would have enough of an influence to push back on unwanted technical control of their brains.

Unfortunately, the middle class throughout the world already has shrunk dramatically over the last 50 years. Whether measured by income or by the amount of wealth accumulated, the gap between rich and poor is substantial and continues to grow, and the group in the middle continues to dwindle. The Pew Research Center notes that, over the last 40 years, the upper-income tier increased from 14 percent of the population to 20 percent of the population. The lower-income tier increased from 25 percent to

29 percent (Horowitz, Igielnik, & Kochhar, 2020). Conversely, with the rich getting richer and expanding, and with the poor getting poorer and expanding, the middle class is being squeezed out of existence, with the middle-income tier decreasing from 61 percent of all income earners to 51 percent. During that same 40-year period, the portion of money earned going to middle-class homes dropped from 62 percent to 43 percent. And the trend continues.

If one considers wealth (defined as savings and one's home minus one's unpaid debt), the middle class isn't faring much better there, either. Adjusted for inflation, the average household in the United States has less wealth than it did 20 years ago (Horowitz et al., 2020). As with the income divide, the wealth divide between the rich and the middle and lower class continues to widen, with the middle class—as defined by wealth—becoming gradually but noticeably closer to extinction.

The dwindling of the middle class isn't just a United States phenomenon. The internationally funded Organization for Economic Cooperation and Development (OECD) conducts extensive research on the topic. The organization defines *middle class* as a household making between 75 percent and 200 percent of the median income for any particular nation. Following that definition, middle-class households throughout the world declined in number from 64 percent of all households to 61 percent of all households during a 30-year period ending around 2015 (OECD, 2019). The decrease becomes more pronounced when observed by age group. "The middle-income group," notes the OECD, "has grown smaller with each successive generation: 70% of the baby boomers were part of the middle class in their twenties, compared with 60% of the millennials" (2019).

Finally, the OCED offers a chilling list of what disappears in the world when the middle class becomes weak or extinct:

- Hope for the future
- Hope that your children will do better than you have done
- The calming sense of stability
- Community investment in education
- Community investment in healthcare

- Community investment in infrastructure and public services
- A disapproval of corrupt behavior
- A trust in your fellow human

While much of this list is hard to measure, the macro-economic contributions of the middle class are very quantifiable. Economists tend to agree that fiscal growth stems from the middle class, as they are the predictable source of purchasers. *Democracy Journal* also agrees, suggesting that the middle class, by its very existence, encourages the positive conditions which contribute to a solid, free-market economy.

> [A strong middle class] is a prerequisite for robust entrepreneurship and innovation, a source of trust that greases social interactions and reduces transaction costs, a bastion of civic engagement that produces better governance, and a promoter of education and other long-term investments. ("Growth and the Middle Class," 2011)

Conversely, when good civics, economic stability, and a trust in the future disappear, the circumstances are ideal for people to embrace an alternative reality and a false sense of belongingness, courtesy of technology inserted in our skulls. The super-rich will be more than ready to provide the hardware.

Incidentally, speaking of the elite versus the masses, there is a school of thought out there that, one day, there will be a dwindling few with access to the Web and to brain-interface technology. (This is the scenario in Kurt Vonnegut's novel, *Player Piano*, where only the elite can access technology, as self-running machines do most of society's work themselves.) The argument goes that brainphones will be a privilege of world citizenship and a way for the elite to control the multitudes. As I alluded to earlier, I tend to disagree. I counter that people tend to buy into whatever types of systems control things, if there is at least a perceived chance that they might "get theirs" someday. And part of the way to perceive such a chance is to have access to the system. A modern example is the stock market.

The brainphone will give us a false sense of equal opportunity. Photograph by Matt Lee. Courtesy of the photographer via Unsplash.

Theoretically, anyone in the world can access the world stock markets, make some intelligent investments, and dramatically improve his or her status. It is a supposition that, if only we had the smarts and a couple of lucky breaks, we could advance. By everyone having a brainphone, we will perceive that we hold the same access as those who actually will be running things—providing us a soothing misperception of equal opportunity and an erroneous feeling of entry.

We Will Love the Ease of Control

Do you like meeting people online and possibly even dating some of them? Then you undoubtedly are familiar with Tinder, the long-time dating app where you are able to look at someone's photo, read a little bit of biography, and then swipe the photo to the right if you think you might like the person, or swipe the photo to the left if you'd like to move on to the next photo. The app is extremely popular, registering billions of swipes every week, and pulling in, at this writing, over $1 billion in earnings a year (Carman, 2020). It is a dating application wonderful for its ease of control.

However, there is a tradeoff for this ease of control—the collection and storage of your every whim, like, desire, and decision revealed on the site. In 2017, French freelance journalist, Judith Duportail, a regular user of the app, decided to ask Tinder for a copy of her personal data, which should have been readily available to her under the European Union's relatively new (at the time) General Data Protection Regulation. With the help of a few privacy activists and attorneys, she pushed a little harder, and Tinder sent her all the data they had on her—all 800 pages' worth! She was amazed at what information they had collected on her, such as the ethnicities of the men whom she had "matched." Tinder also had gathered data having little to do with her activity on the Tinder application itself, such as how many "likes" she had offered on her Facebook account and where she had electronically stored her Instagram photos.

Perhaps more astonishing is that Duportail seemed to indicate in her article (published by *The Guardian* and later turned into a book, *Love under Algorithm*) that she hasn't stopped using the Tinder app! "As a typical millennial constantly glued to my phone, my virtual life has fully merged with my real life," she writes. "There is no difference anymore. Tinder is how I meet people, so this is my reality. It is a reality that is constantly being shaped by others—but good luck trying to find out how" (Duportail, 2017).

So, here's a company that most people have no direct, important, or physical contact with, and yet that company manages, through asking a few questions and watching the actions of its app users over time, to garner thousands and thousands of data points on millions and millions people. More dismaying, people pretty much know the company is collecting this information (at least these days they do), and yet they continue to use the app. Surely, there must be a reason other than apathy and loneliness for people to allow so much personal information to be gathered (and maybe eventually hacked and/or sold) about themselves.

Perhaps the reason we shrug and allow corporations to conduct this type of espionage is no more complicated than the appeal of convenience. Tinder, as mentioned, offers a basic concept—swipe right to like the image and swipe left if you're not interested. The user feels very much in charge, and the ease of control is so inviting.

There are many things about the brainphone I can argue against, but the one thing I can't dispute is that the brainphone will create convenience in ways we can't yet imagine. The notion of locating anyone in the world and communicating with them just by thinking about them is pretty wild. Calling up old memories and reliving them in detail, creating false memories (from a catalog?) and living them for the first time, living someone else's memories—the possibilities are endless and the virtual scenarios likely will be very, very entertaining. (I don't plan on getting the brainphone, but if I were to get one, I'd knock about 40 pounds off myself as an adjustment/tailoring to all of my middle-aged memories.)

Ordering groceries or clothing and paying for them with just a couple of distinct thoughts; checking your medical records and exercise logs whenever the thought comes to mind; arranging a first date with a stranger; instantly gaining any new knowledge or ability, such as knowing how to play an instrument; holding a virtual meeting with your work team in your mind, with the half-dozen of you in half a dozen different places; video gaming in scenarios that look and feel extremely real; cosplay, not by wearing a costume but by actually assuming the look and sound of the comic-book character—wow, wow, wow. All low-hanging fruit for the picking. All extremely convenient.

Before asking, "But at what cost?" and "But to what end?", here's a more basic question: Won't it get tiresome relatively quickly? It reminds me a bit of the stories of people who win the lottery, only to grow weary of the money and find that it doesn't lead to happiness. Okay, so what about the simulated sex a brainphone might offer? After all, there are very few things in life more desirable than procreating, and a near-perfect simulation (aside from being harmless and disease-free) couldn't possibly become boring, could it? I suggest that it absolutely could. Rock legend Sammy Hagar often was asked about all the women he was surrounded by in his early days of touring. His response, with a shrug: "You know, even when they're nude and for the taking, it all looks the same after a while." Maybe being able to know anything, go anywhere (virtually), and do anything (virtually) are ultimately unfulfilling and sadly numbing.

Okay, so back to the first two questions. At what cost? And to what end? If the cost of ultimate convenience is allowing corporations and governments to know everything there is to know about you and to pass that information around with few boundaries, I would argue that scenario will lead to very bad things. Joshua Becker, writer and advocate for a minimalist lifestyle, makes the argument that each uptick in convenience has a responding downtick that impacts some aspect of our lives. The convenience of having prepackaged food delivered to you, for example, distresses your health, because such food tends to have way too much salt in it. The convenience of having things we want instantly impacts our ability to sit back, be mindful, meditate on what's really important, and strategize long-term ways for achieving real happiness. Becker also laments that a society with uber-convenience is a society less challenged and, therefore, less likely to learn and improve. "Often times, the greatest lessons we learn in life are born from inconvenience, or pain and suffering," he points out. "But among a society where convenience and comfort are pursued above everything else, the

Minimalist Joshua Becker says that each uptick in convenience has a responding downtick in life quality. Photograph by Gabriella Hileman. Courtesy of Joshua Becker.

Our love of convenience might be what leads to our destruction.

opportunity to develop perseverance becomes less and less frequent" (Becker, 2016).

And, finally, to what end? I suspect you know my answer already. The end to our accelerating pursuit of convenience is the end of the human race, or at least the human race as we recognize it. Loving the convenience to the point where we cannot turn away from something we *know* is going to harm us or destroy us is an unfortunate possibility. The likelihood of this phenomenon happening in our lifetimes, or even within the next several years, is high, and it is dreadful. The good news, however, is that this future is not certain. We can plot a different course for our existence. Hence, the reason you are reading this book.

Remember, corporations and governments do not exist to help you. They are here to take advantage of you, control you when they deem necessary, and ultimately to harm you. It is the side effect of the power they can't get enough of. It's not their fault—they're addicted.

PART TWO:

THE MEANS TO CONTROL US

CHAPTER 4:
HOW WILL YOUR BRAINPHONE WORK?

And thus I saw the horses in the vision: those who sat on them had breastplates of fiery red, hyacinth blue, and sulfur yellow; and the heads of the horses were like the heads of lions, and out of their mouths came fire, smoke, and brimstone. By these three plagues a third of mankind was killed—by the fire and the smoke and the brimstone which came out of their mouths.
(Revelation 9: 17-18 NKJV)

Your Brainphone Will Be Easily Implanted

Inventing a device that could break down people's thoughts or dreams into computer bytes and allow other people to watch them on a screen always has been a fantasy of scientists and science-fiction enthusiasts alike. I remember being blown away as a young man by the Joseph Ruben movie *Dreamscape*, where you could enter someone's erotic dream (and, well, participate if you'd like) or someone's nightmare (and either fight the monster in the dream and help the dreamer, or become the monster yourself and scare the dreamer into a heart attack and an early grave). Dennis Quaid was so cool. Heck, he still is!

Coming up with a brain-tech interface tiny enough that it can be implanted, unseen, in the human skull is, in many ways, just as fascinating. The miniaturization of processors and memory chips, which began in the late 1950s, has over the last decade been gaining speed. As a result, there is literally a thousand times more computing power in today's tiny smartphone than there was in a 1970s supercomputer that used to take up an entire office of space. Now, we are at the dawn of a new phase of miniaturization, where the technologies that were shrunk down in electronics and computers, in turn, are being used to discover ways to shrink down mechanical devices. In fact, such technologies soon are going to be able to copy and shrink down just about any three-dimensional object. There are some laws of physics that need to be addressed, but microscopic motors and such are essentially here.

The current plan for brainphone implantation includes a small, augured hole in the skull, just beneath the skin, to house the device, with wiring so small that the brain tissue is oblivious to it. With miniaturization at its current pace, the tiny device likely will be placed by robots directly into the part of the brain which houses the hub of most cognition.

Speaking of robots: In Chapter One, I mentioned that the robotics used to insert the brainphone will be mass produced, making an implantation center (brainphone store?) just as easy to find as, say, a LASIK center or, perhaps, an ear-piercing pagoda. When my daughters (in their early 30s) read an early version of this manuscript, they were incredulous about the notion of mass-produced robots conducting these procedures. "Sorry, dad," said one of them, "but that will never happen." So, a near-future device, implanted in your mind, creating new knowledge instantly and saving old memories, is conceivable, but a robot conducting the implanting procedure by itself is ludicrous?

A robot will be at your local mall to do the implanting. Photograph by Yuyeung Lau. Courtesy of the photographer via Unsplash.

The funniest part of their skepticism is that, at this writing, the surgical robot that implants the device is one aspect about the brainphone that *is* real, right now. Realizing that a human couldn't handle the intricate threads connecting their *link* to the brain, Neuralink put the designing and building of a robot that could handle the super-precise sawing and sewing front and center. The surgical robot is now complete, and it has successfully implanted the *link* in pigs and monkeys. Since the robot is able to avoid blood vessels in the brain that a human micro-surgeon cannot, I suspect that these robots will be used in future brain surgeries that don't even involve the brainphone. Whether or not a patient goes to the local shopping center, where the nearest robot is housed, to have a brain tumor removed remains to be seen. I hope the store pays its electric bills!

So, as currently planned, Elon Musk's *link* process will be a one-hour, out-patient procedure, using a local anesthetic (which means you'll be awake while the robot does its thing!). After the operation is complete, you'll be able to leave. As with everything, the process is likely to become easier and more affordable over time. Just as LASIK surgery used to be lengthy, painful, and expensive and is now fast, painless, and reasonably priced, so, too, will be the practice of receiving a brain-tech implant.

In Chapter One, I also noted that an older generation of consumers might be more than hesitant in purchasing a brainphone. My guess was that some debilitating illness would then befall the human race, and the brain rewiring capabilities of the brainphone would save the day. Older folks, previously opposed, would climb on board the brain-tech-interface train. But what if there's another way to gain the liking of the age group with the second strongest purchasing power? (Baby boomers recently have been overtaken by millennials both in population size and in per capita ability to spend.) Suppose the willingness to drastically alter oneself at an older age happens when the ease of surgery is combined with the ease of older living. A good example is sleeping. Older people have a more difficult time falling asleep and staying asleep throughout the night. Also related to sleep: older people are hit a bit harder with jet lag after taking a long flight. Gabe Newell, founder and CEO of Valve Software, suggests that brain tech will not only be able to place people in virtual realities, but it might be able to place people in dream states. That is, your brainphone might be able to help you sleep better. Says Newell:

> Sleep will become an app that you run…it's like, I just say, "This is how I want to sleep right now."…Most people will look at that and go, "That sounds pretty nice. You mean I got on this twelve-hour plane and I basically said, 'I do not want to be awake for the next twelve hours,' and I wake up completely refreshed with my circadian rhythm?" People will say, "Okay, this other stuff sounds scary…[but] I'm going to go ahead and try that." (Appleby, 2021)

Hmmm, I wonder if a brainphone could help me, as an older man, not have to go to the bathroom every 15 minutes! Where was I? Oh yeah, Newell's predictions. Okay, so let's take Newell's example one step further. Suppose you're about to get on a long plane ride, you're worried about jet lag, *plus* you're afraid of flying. Now suppose there's a brainphone kiosk inside the airport terminal, and you have an hour or two to kill? (Assume that the recovery from receiving an implant isn't affected by the change in air pressure on a plane.) The ease of the procedure and the instant gratification you will get when your plane lands might be too much for you to resist.

In sum, there will be all kinds of reasons that people might shy away from the brainphone, at least initially. But the ease of implanting, along with the near-instant convenience, isn't one of them.

Your Thoughts Will Be Converted into Code

Phonograph, long-playing (LP) records have made a small comeback in the world. People seem to like the sound of analog recordings, scratchy though they might be, as the record-player needle touches the vinyl. When people claim the sound is superior, I don't think any of us really believes them. What they *really* mean to say is, "I prefer this recording because I know it's *real*." That is, the recording hasn't been broken down into computer code, transmitted, and then restructured back into something resembling the original sound. When we hear Charlie Parker play the sax on an old vinyl, we *know* it's him.

The comparison is unfair, to be sure. Digital recording and digital converting have improved dramatically over the last 30 years. Plus, once the sound has been broken down into a series of numbers, it can be stored and duplicated indefinitely without degrading, making it far superior from a posterity standpoint. However, it still doesn't feel as authentic as the inferior, analogue recording. Why do we have such a response? Is it simply romanticism? I'll go along with that notion: I'm an old romantic. But I think there's something more—a sense within us that digital is not reality, and, as humans, we inherently prefer the analog. The *real* thing. We don't know why we feel it. In fact, we might not even *notice* that we're feeling it. We just do.

That is one aspect of the brainphone we should find most concerning: In order for it to function, it must break down our thoughts into code; have them transmitted (and perhaps stored) as code; and then, at some later point, reconstruct the code; and have the results considered as the original thoughts. There are many, many factors that the coding is not likely to take into account. For example, what were the feelings behind the thought—was the person doing the thinking in a foul mood because of something that had happened earlier that day? What was the context of the thought? If a woman thinks, "Oh, I want to kill that husband of mine," is she simply exasperated with him and exaggerating (maybe even lovingly!), or is she, indeed, planning to murder him in his sleep later that evening? What was the person's alertness? Did he have the thought with all his wits about him, or was he groggy—or even asleep and dreaming? When you magnify all those zeros and ones, you'll see a lot of space between them. Computers have to fill in those gaps or ignore them. It's a nuance, I argue, that even the most advanced AI won't be able to handle. Not to be overly dramatic or romantic, but those spaces between the zeros and ones are what make us humans. If we discount those spaces—that nuance—we basically turn our backs on our humanity.

Another thought about code: it essentially lasts forever unless it is intentionally wiped away or unless there is no more storage space for it. On a computer, specific data can be recalled instantly, with large amounts of data on a particular topic gathered in relatively short amounts of time. When we are physically one with the computer, does this feature become a blessing or a curse (or deadly)?

Some very intelligent people have speculated on what life with near-perfect memory would be like. Their assessment: not good. One of the first people to write about such a thing was, coincidentally, the Father of American Psychology, William James, who suggested (back in 1890) that thinking requires generalization, and that generalization requires at least some amount of habitual and multi-tiered forgetting:

> If we remember everything, we should on most occasions be as ill off as if we remembered nothing…The paradoxical result [is] that one condition of remembering is that we should forget. Without totally forgetting a prodigious number of states of consciousness, and momentarily forgetting a large number, we could not remember at all. (James, 1890)

American psychologist Williams James. Public domain photo. Courtesy of Wikimedia Commons.

In other words, if we don't allow our minds to forget—at least temporarily—the huge amounts of what we've learned and experienced in the past, we won't be able to make the types of conclusions that allow us to function each day.

I found James's great quotation in an article about the perils of never forgetting, written by Rodrigo Quian Quiroga, who, at this writing, is Director of the Center for Systems Neuroscience at the University of Leicester (England). Quiroga comments on "Funes the Memorious," a 1944 short story about a teenage who injures his head after falling off a horse. He awakes with the ability to remember every minuscule fact of everything that he has ever read or seen. Rather than benefit from the talent, Funes seems to wither from it, and, in the end, the condition seems to figuratively swallow him up and kill him at a very young age. In his literary review of the story, Quiroga obviously leans towards the curse side of such a talent over its blessing side. He compares being able to instantly remember "vast labyrinths of memory" to "[playing] with the infinite," and warns of "the consequences of having an unlimited capacity to remember" (Quiroga, 2019).

Incidentally, up until about ten years ago, most brain experts believed that forgetting was an unplanned brain process, brought on by the gradual erosion of memories that were old or not being used enough. Then, a few groups of cognitive researchers began exploring the possibility that forgetting was a more active, strategic process. That is, when it came to forgetting, perhaps there was a method to the brain's madness. In fact, the brain's initial response to input—its default mechanism—is to forget it rather than to remember it. Something else has to happen for the brain to switch over to its customized setting, which is to remember and store the data. Canadian cognitive psychologist Oliver Hardt suggests that you cannot process data and save it unless you forget much of it. "What is memory without forgetting?" he asks rhetorically. "It's impossible. To have proper memory function, you have to have forgetting" (Gravitz, 2019). To put it bluntly, evolution has caused the brain to recognize that too much information is not a good thing.

University of Toronto Scarborough neural scientist Blake Richards goes even further by proposing that the brain uses tactical forgetting as a tool for survival. Richards offers the example of being bitten by a dog. What if you were to remember every little detail about the experience? The weather that day, the breed of dog, whether or not the dog barked, whether or not the lawn was mowed, the perfume that the dog's owner was wearing, etc. In such a case, it might be difficult to pull out the important facts that could help you from being bitten again down the road. But if your mind washes away the minutia and the unimportant items, it can retain and piece together the things (such as spotting a snarling dog off its leash) that might cause you to turn around or go inside. Notes Richards: "If you wash out a few details but retain the gist, it helps you to use it in novel situations. It's entirely possible that our brain engages in a bit of controlled forgetting in order to prevent us from overfitting to our experiences" (Gravitz, 2019). While it is certainly possible that brainphones in the very distant future might teach themselves to strategically forget, it's highly unlikely the first couple versions of the device will be able to do so. And humans might pay a hefty price by the deluge of facts they are unable to wade through.

The converting of human thoughts into code might make for ease of storage and transmission. But important things might get lost in translation. And people might get drowned in data.

You Will Search and Click Just by Thinking

At this writing, there is a big drawback to brainphone technology: The wires that connect the device to the brain are limited in their scope. That is, if you want your brainphone to monitor or enhance your motor functions, then the wires from the implant must go to the specific part of the brain that controls the movement of your limbs. If you want your brainphone to install a new language that you instantly can understand and speak, then the wires must join the part of the brain that runs language. And so, for the advanced types of things we're expecting these implanted upgrades to do, we will need to have hundreds of thousands of these near-microscopic wires running out of the brain-interface technology. The set-up might be feasible, even at present, but people are less likely to go along for the ride if the wiring is so extensive and overwhelming.

Here's what I see happening very soon. Outer space is vast, in ways very difficult for us to comprehend. Even if we designed a spaceship that could travel at the speed of light, it would take over four years just to reach the Proxima Centauri, the nearest star to our sun. So, in our theoretical writing, we have come up with a way around the great distances—find two wormholes, go into one, pop out the other one, and, ta-dah, we have arrived somewhere on the other side of the universe. (It is, in fact, the same wormhole on both ends, but that's another story.) Einstein's math suggests it might work someday, and we might have complex maps of which wormhole openings lead to wherever.

Okay, so now apply the same concept to *inner* space. I suspect that connecting one end of the brain, with wiring, to the other end of the brain is much the same as traveling in a straight line from one star to another: it's possible, but not necessarily practical. But what if there is a portal in the brain that leads to everything else? I suspect it is there. I also suspect that people way more intelligent than I am know it's there and currently are looking for it. Once they discover it (which likely is soon, due to the hundreds of millions of dollars in brain-tech research going on right now), then technicians simply will have to hook up the brainphone to this newly discovered brain wormhole, and a full-immersion experience will be possible.

Once the brainphone/wormhole interface is in place, your brain will work very much like a computer screen. You will "see" the screen, even though it really isn't there, and you will be able to see a cursor and move it, just by thinking it to be so. You will see drop-down menus that can take you to various memories stored in various regions of your brain. You also will be able to run a diagnostic test on both your brainphone and your brain whenever you want to, just by thinking it to be so. (Courtesy reminder: Make sure your brainphone is fully charged before running these types of checks.)

You also will see drop-down menus for searching for information; going onto social media sites; making purchases; gaming; instantly learning new information; listening to music; watching movies; and entering alternate realities. (More on that last one in a minute.) In other words, all of the things you currently are using your computer for will be available—with a mind click—in your head, powered by the amazing processor that is the human brain and augmented by a hyped-up human

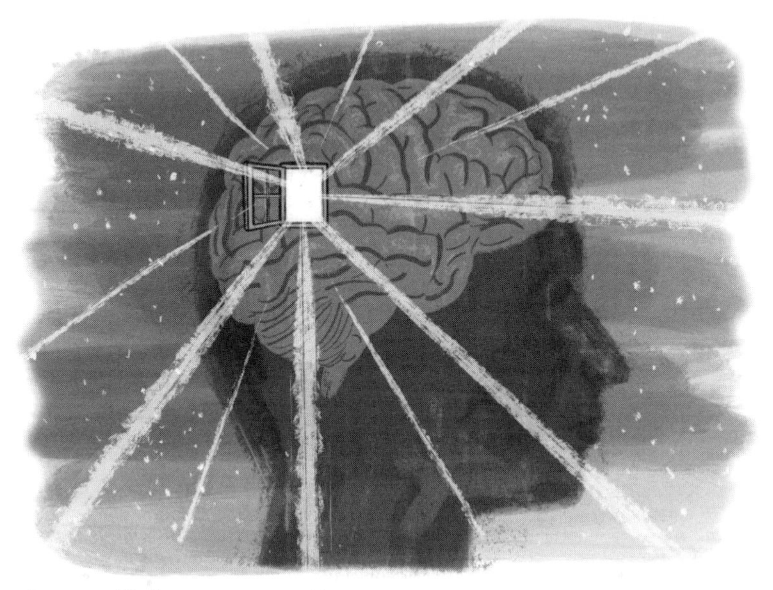

Soon we'll discover a portal in the brain that leads to everything else.

imagination.

I wouldn't say that smartphones drove me out of full-time university teaching. Truth be told, it was probably the rotten pay. But the existence of smartphones surely made leaving academia (for, ironically, a job in IT) a lot easier. How disheartening it had become, year after year, to teach a lecture hall full of students, with, say, 70 percent of them looking at their phones. One nice thing: At least some of them were courteous enough to have their laptops open, so I could lie to myself that they were taking copious notes. God bless them. It was, of course, much more likely that they were gaming or IM'ing or checking out their likes on social media. I tried to foil the phones by asking lots of questions, playing trivia games for quiz points, and keeping the topics controversial and the discussions lively. The results were lukewarm at best: people just wanted to be on the Web. I can't imagine what in-person teaching will be like soon, with all students having their computers and phone "screens" embedded in their minds. A student looking right at you and nodding his head might be listening, or he might be watching a music video. Or he might be watching porn! Who will know?

This book already has discussed, at length, addiction to smartphones. But what about addiction to the Internet in general, regardless of the device? As with smartphone addiction, the tendency of staying on the Internet to habit-forming levels is most common among teens. Craving the Internet to the point of no longer functioning well in society is a very real thing. It can involve addiction to gaming, pornography, shopping, or gambling, none of which inherently requires a smartphone. Admittedly, none of these things requires the Web, either. Gaming is available at an arcade. Pornography is available on pay-per-view. Shopping is everywhere. And gambling can happen in a casino if you're old enough to walk in or in a basement card game if you're not. But calling up any of these things onto a screen is just too, too easy, making it difficult for the addict trying to avoid "people, places, and things." Now apply the brainphone to the situation. Imagine magnifying the intensity of these addictive things by 100, and reducing the ability to walk away down to zero. Whose life *won't* be ruined by overexposure to these habit-forming stimuli?

Once you can search and click in your mind, your outer real world will become increasingly less important. Concentrating on what's in your mind screen—and tapping into constant exhilaration—will take up much of your time and brain power. Remember, you can search and click now just by thinking.

Your Real World and Virtual World Will Feel the Same

Imagine waking up one morning in a strange bed. You feel very much at peace. You can hear birds chirping outside, and sun beams are pushing their way through the sides of the curtains. There's an empty bottle of wine and a couple of glasses on the nightstand. You still can smell the wine's lingering, fruity aroma. You sit up, rub your eyes, and look next to you. Lying naked, partially covered by the sheets, and still asleep, is the most beautiful person you've ever seen. The day apparently aims to be as pleasant as the one before.

Then you have a thought. "Is this real, or am I still in a brainphone simulation?" The funny thing is, you really don't know.

You might look over at the warm body next to you and have another thought: "Should I care?"

But, of course, you have to care at *some* point. You might have to check in with family members to make sure they're okay. You might have bills to pay or an important meeting at work coming up. You might have medicine to take. For goodness' sake, if the situation you're in right now is virtual instead of real-life, you have to drink some water in your other reality, or your real, physical self will die from dehydration! (At some point, you also might consider whether or not you're cheating on your spouse, even if it's not real. Hint: You probably are.)

When our thoughts are broken down into code, creating simulations for us to enjoy will be particularly easy for a company. Much of what will exist in our fantasy worlds will be of our own, dream-based conception—the programmer simply will be helping it along. My guess as to how we'll know if it's real or not: In the lower, left corner of our peripheral vision, we'll see a capital R or a capital V floating in front of us, telling us at any given time if it's live or if it's Memorex. (Reference to an old TV commercial. You'll have to check it out.)

In the action-packed 1999 movie *eXistenZ*, sci-fi master David Cronenberg tells the story of a video game so real that the participants, at some point, no longer know if they're in the game or back in the real world. Two gamers, played by actors Jennifer Jason Leigh and Jude Law, are so traumatized by the experience that (spoiler alert), at the end of the movie, they assassinate the game designer. Wow, seems a little rough, right? After all, no one ever shot the inventor of the rollercoaster for creating a ride that made them lose their cookies. In the movie, after shooting the game designer, they point their guns at another player who allegedly has ties to the designer. He utters the film's final line, which leaves the movie-goer hanging: "Hey, tell me the truth—are we still in the game?" Since he's chuckling slightly, he genuinely doesn't seem to know. And neither do we.

There's little doubt that playing video games in realistic, virtual worlds tends to make young people more aggressive. A

2014 meta-analysis of nearly 100 different psychological studies on the topic (observing and surveying a total of 37,000 gamers) credibly argues that violent video gaming produces more violent kids (Greitemeyer & Mügge, 2014). Whether or not full-immersion, brainphone gaming will intensify the effect is enough of a topic to fill another book.

But science fiction stories like *eXistenZ* (a pretty insightful movie considering it was released before the turn of the millennium) put forth a few important questions regarding not knowing what reality you're in, as it relates to violence outside of gaming. First, does not always knowing what reality you're in make you less sensitive to violence? If you practice, say, martial arts in a virtual world, fighting to the bloody death, does it desensitize you when you go back into the real world? Are you more likely to use the skills that you learned and bled through with your virtual Sensei? Are you numbed by it all?

Second and conversely, does not always knowing what reality you're in make you hypersensitive—not to the violence, but in your moods? As mentioned earlier, part of the problem with smartphones and social media is that they have crafted a generation of extremely thin-skinned young people. Many of them are touchy all of the time, and each of their days often involves running the gamut of emotions. If you're punchy as it is, what happens when the device and the electronic posts become physically intertwined with you? Furthermore, does not knowing where you're at contribute to the hypersensitivity? During my time the U.S. Army, I attended jungle warfare school at the Jungle Operations Training Center in Panama. The heat, the snakes, the threat of debilitating crotch rot—none of those things scared me. But when I was in the jungle at night with a map (but no smartphone or global positioning device, as neither had been invented yet) and no good sense of my location, I was pretty afraid. "Where in the heck are we headed, Snair?" "I'm not sure, brother, but we're making great time!" It's hard to know where you're going if you don't know where you're at, both in geography and in life. Along with the possibility that the real/virtual vagueness might numb us, there's the possibility that the ambiguity might unnerve and frighten us.

Third and finally, if—at any given time—you don't know where you are, are you apt to commit the occasional, violent act to find out? Sort of a "pinch me so I can tell if I'm awake" act, only more gruesome. If the two worlds are indistinguishable in every way but the pain (due to selective programming), then a bit of violence that hurts might help orient you. ("Ouch, that hurt! I guess this is real.") If, on the other hand, pain exists in *both* worlds, then perhaps the act might be psychologically grounding, the way troubled young people sometimes cut themselves to feel steady. Remember, aside from waking up their emotional numbness or using pain as a sense of stress relief, some teens cut themselves because, unlike the vagueness of anxiety or the unknown future, self-inflicted physical pain is real and specific. Notes Johns Hopkins All Children's Hospital:

> Emotional pain can feel vague and hard to pinpoint, talk about, or soothe. When they cut [themselves], teens say there is a sense of control and relief to see and know where the specific pain is coming from and a sense of soothing when it stops. Cutting can symbolize inner pain that might not have been verbalized, confided, acknowledged, or healed. And

Will you have to hurt yourself to see if it's real or virtual?

> because it's self-inflicted, it is pain the teen controls. ("Cutting," 2021)

How sad. So, if your brainphone creates disorientation, will you hurt yourself (both in your real world and in your virtual world) in an attempt to anchor yourself?

If the real/virtual uncertainty presents a problem for the individual brainphone customer, how will this confusion project out onto society when millions of people are going through life not quite knowing where they are? What if even a small percentage of them neither knows nor cares? And, even if the distinction—at some point—becomes relatively user-friendly, how will society function when everyone is muddled and preoccupied with the last (or the upcoming) utopian simulation? It is undoubtedly quite a chore to sit down and pay your monthly bills when, at any given moment, there's an imaginary lover waiting in your imaginary bedroom (courtesy of The Corporation). How will society function? My argument: It is likely to *not* function. The word *mayhem* comes to mind.

As you might have guessed by this point, my biggest fear about the brainphone is that it will inevitably cause us all, collectively, to cease functioning in any meaningful way as humans. My second biggest fear is that, as we lose track of what is real and what is digital, we simply will settle in to our virtual worlds, so content within our fantasies that we eventually die (in real life) of thirst and starvation.

Social Media Will Be a Part of Your Daily Thought Process

I hate to disparage social networking. On its surface, having a group of friendly contacts online seems harmless. For the shy person who might otherwise avoid people, it might be a nice way of learning to socialize. For an elderly person who can't get out much, it might be the daily break from isolation. And for a paraplegic person, social media might be the only alternative to being a prisoner of one's home.

During the coronavirus pandemic, social media took on a whole new purpose—keeping people connected while they stayed physically separated (to slow down the spread of the virus). While there was a lot of conflicting information going out, at least it was accessible. And it went out immediately. Also, meeting sites, education sites, and other sites similar to social media took the place of classrooms as the primary way to teach K-through-16 students. Finally, if it wasn't for the Web, my family and I would have gone for the better part of a year without "attending" church. If it wasn't for social media sites, small prayer groups would not have been able to cover their weekly *Holy Bible* readings and discussion topics.

Social media sites give us access to other cultures and the opinions of people throughout the Free World. They allow researchers throughout the world immediately to share the results of scientific or academic research and have similarly-educated world colleagues examine the data and comment on it.

Remember, the Internet was designed by U.S. government researchers as a way for the U.S. Defense Department to communicate after a nuclear war. The idea was that, with a "net" or "web" of individual communicating stations instead of one central station, a huge part of the system could be destroyed and the system still could function. And so, what better fit is there for such a structure than a web of friends forming a distant-but-intimate community? That is, the Internet was, in a way, designed specifically for social media.

I'm not sure when, exactly, people started to figure out that social media was potentially harmful to users, especially young people. Lukainoff and Haidt (2018), mentioned in Chapter Two, specifically mark the starting point as October 2013, when Facebook began allowing teenagers age 13 to 17 to post publicly, the same as users 18 and older. The year 2013, as it turns out, was also the year that smartphone use went from being introduced (six years prior) to being a staple of American teen culture. The combination of the two (social media and smartphones) made for what turned out to be a new generation of young adults who are extremely sensitive, fearful, lonely, and more prone to hurting themselves or committing suicide. Lukainoff and Haidt comment: "In short, iGen [young Americans born between 1994 and 2007] is the first generation that spent (and is now spending) its formative teen years in the giant social and commercial experiment of social media. What could go wrong?"

I remember the first time one of my daughters cried when someone posted something derogatory about her on Facebook. She was really shaken up. It was an old boyfriend that she had forgotten to "unfriend," and he felt like raining on her parade as she announced some good news. He didn't live in the area anymore—so I couldn't drive over and kick his butt. But I don't think I would have done so, anyway. I'm a lover, not a fighter. Besides, quite frankly, I didn't think what he wrote was all that disparaging. Obnoxious, yes. But offensive? Not really.

My advice to her at the time was, "Forget about it. Facebook is not reality. It's just words on a screen. And the words on the screen are just electrons swirling around in a plastic box. It's not real."

These days, I'm not so sure that was good advice, or that it would be in today's terms. Children are, after all, traumatized by what belittling things are written about them on the Web, and cyber-bullying seems very real, especially if it results in poor mental health or suicide. French reporter Judith Duportail put it best in the quotation used in Chapter Three: "My virtual life has fully merged with my real life. There is no difference anymore." The point of this book is not to convince people to put down their smartphones—that would be a fool's errand. Instead, the goal is merely to convince people not to have smartphone-like devices sewn in their heads. However, when you hear a young, bright person in this world say, "My apps are my life now," it becomes evident how difficult the goal might be. Are we already too late?

At this writing, I do not have a personal Facebook account. I tried it out a dozen years ago, about four or five years after it had been around. It immediately began taking up large chunks of my day. Plus, it struck me as dreadfully habit-forming. I broke away and cancelled my account before I got too sucked in. But what happens when Facebook, or something like it, becomes part of our daily routine with our brainphones plugged in? That is, what happens when we *need* Facebook to function in society? What happens when addiction becomes, well, an essential part of our existence?

The question is more than existential. There are some very practical concerns about the communication features of the brainphone. For example, Elon Musk has suggested that people with his *link* will be able to communicate without speaking—telepathically, so to speak. Also, we will be able to send ideas to each other as concepts, rather than as statements. People will share concepts as groups. Hmmm, have I used the word "hive" elsewhere in this book? It sure seems applicable here. People strike me as very uptight these days, in part for reasons related directly to today's technology (as covered in Chapter Two). How uptight will people become when they have to rely on communication-by-concept, or by mutually-shared or mutually-approved concepts, to get a point across? If nothing else, I'm guessing there will be a lot of slander/libel/harassment lawsuits, when people feel their expressions were telepathically mis-stated and misconstrued by The Hive.

Just as the negative attributes of the smartphone will be magnified once the device is implanted, so, too, will the negative attributes of social media be magnified once the media is intertwined with our minds. For example, social media is considered partly responsible for communal "echo chambers," where a person joins a like-minded group with the same political and/or religious outlook. The opinions, since they are similar, bounce around the Web like an echo, and the person believes what he wants to believe and hears what he wants to hear, validated by people bouncing back the same ideas. If someone finds her way into the virtual group and proceeds to offer different or opposing views, the group becomes hostile, and she is summarily tossed out. Facts are what the group deems them to be. Truth is what the group deems it to be. External truth matters not. Now, imagine, with brainphone wiring, that you have to live your life partly within the realm of this social-media reality. If your group collectively has decided that an apple is, in fact, an orange, how are you supposed to go into a supermarket and ask where the apples are?

I once taught young K-through-12 teachers how to look for employment once they received their teaching licenses. The first bullet point on my list was to cancel their social media accounts during their interview phase and, if possible, during their probationary phases (usually lasting between one and three years). Most of them laughed. But I handed out a list of reasons—taken from recent news stories—why new teachers had been fired or why would-be teachers had had employment offers revoked. They included some dumb things, such as "friending" students. But they also included some things that a reasonable person likely wouldn't find offensive. One young teacher posted a photo of herself drinking beer with some friends (all of legal drinking age). Another posted some photos of herself modeling bikinis for an online summer wear boutique. Another posted his thoughts that one of his U.S. senators was a jerk. None of these postings, incidentally, had been fully public. They had been posted for a network of so-called friends. In each case, someone (either because of pettiness or the pleasure some people find in ruining others' lives) passed on the photos or the comments to school officials. My suggestion to my classroom of future educators was that they not only be extremely careful about what they posted on their social media accounts, but that perhaps they should "keep a low profile" until they had become somewhat grounded in a school system—enough that, if students and administrators appreciated them, they might be able to bounce back from an embarrassing Web episode in the future.

So, how would my advice apply in a world where social media, via brainphones, was the manner in which everyone functioned? In such a world, while you were speaking to a potential employer, she could look up social-media photos of you while you were sitting in front her, still answering questions, during a job interview. Before the screening was finished, she might know all about your friends, your habits, your evening hang-outs, your political views, how well you spell words on your media, how often you use profanity while chatting online with friends, your online dating preferences, and if anything had ever happened to you that was controversial, as discussed on a social platform by others *besides* yourself! In an existence where social media is magnified and obligatory, maybe your reputation could be ruined before you even knew you *had* a reputation, with very few ways to remedy the condition. When people find old, digital relics of your existence instantly—always there, always to be humiliated by—you will need to have quite the sense of self to continue on. A good sense of humor might help, too.

This brain-tech interface device will really be something. It will be implanted quickly and painlessly. Your thoughts will be

During a job interview, the employer might be looking at your social media history on her brainphone, even as you're answering her questions. Photograph by Tim Gouw. Courtesy of the photographer via Unsplash.

easily transmitted and stored—forever, if you'd like. You will search for information throughout the world and find it, just by thinking. Fantasy worlds will seem so real that you'll have trouble noticing their illusion. And your social-media friends will be in your head, and will be your everything. How will your brainphone work? Amazingly. That's what's so frightening about it.

CHAPTER 5:
HOW WILL THE NETWORK BE MAINTAINED?

All things are lawful for me, but all things are not helpful. All things are lawful for me, but I will not be brought under the power of any.
(1 Corinthians 6: 12 NKJV)

Elon Musk and Others Will Continue to Innovate

When I got out of the U.S. Army about 30 years ago, I had ambitions of becoming a manufacturing manager and, eventually, to teach management (both of which eventually came true). But, with a wife and two young children, I had to put those aspirations on the backburner and start making money right away. And so, I took a job selling mobile phones. It wasn't a bad gig. The commissions were good: back then, the phones were very expensive and often installed permanently in people's cars. At least initially, the only people who could afford them were business people who were on the road a lot, such as plumbers and home contractors.

Anyway, when I was selling those wireless phones, many folks in my industry were wondering when the cellular towers would be replaced with a network of satellites. It was something of a concern for anyone in the business. The thinking was that the first company to launch a network of satellites for wireless phones and the Internet would take over the industry. Everyone waited with bated breath to see which company it would be.

Unbelievably, for 30 years, those same breaths were being held! In a world of accelerating technological advancement, the wireless business surprisingly became comfortable with cellular towers, especially as the networks of towers crossed the country and the expensive "roaming fees" for jumping from network to network while traveling went away. It is only now that the vision of an international network of wireless satellites in space is turning into reality.

And who is it that's seeing this vision through? Elon Musk, of course!

I must admit, aside from his really bad brain-implant idea, I am very much in awe of this man's ingenuity. I never thought, in my lifetime, I would see a rocket return to Earth from space, spin around, and land gracefully on a platform with thrusters a-blazing. Musk's SpaceX rockets can do it. I never thought I'd see people pods traveling in a tunnel at 600 miles per hour, carrying a family from Baltimore (where I grew up) to Washington, DC, in eight minutes. But Musk's *Hyperloop* is in the permit stage, and I think I'll sit in one of those pods someday soon. (I hope I don't spill my coffee!) Musk continues to improve on his vision of electric, self-driving cars. And his *Starlink* network of tens of thousands of Internet satellites is well on its way to being fully launched into orbit.

Once the hardware is up and running, the world will be transformed by one dominant, easy-to-access Internet network, available even to parts of the world that have poor Internet service or none at all. This network also will be controlled by one, very powerful entity.

If Musk's sense of innovation is projected onto the brainphone, I suspect the device will continue to change quickly in ways that I cannot imagine. For example, what if the implant one day is manufactured with living tissue, with organic circuitry and jelly-like memory storage? Might it be implanted, in this organic form, at birth? Your first day in the world, and you get your circumcision and brainphone before you know what hit you! Would people be as opposed to a brainphone that was flesh and nerves like us? And, if people wanted to draw the line between humans and technology, how blurry would that line be if the machinery were organic? The questions go on and on. And, perhaps, Elon Musk has these new, madcap (pardon the pun) ideas already thought up and planned out.

If, as Plato suggested, necessity is the mother of invention, then one has to wonder what is necessary about the brainphone. Some of the medical promises of the device (helping to cure paralysis or dementia) hint at necessity. And, certainly, if entertainment is something of a necessity, then the virtual gaming worlds promised by the device make it a gotta-have by any serious gamer.

Although Musk is quite the showman, and he relies on hype as much as anything else to keep investors interested in his projects, he certainly isn't the only major player that we can expect to see major innovations from. Also investing millions of dollars in brain-machine interface (BMI) technology is Meta (Facebook) CEO Mark Zuckerberg. Zuckerberg is likely to be the most innovative regarding a neural communication network, as his BMI plan reportedly involves less-invasive neural caps that fit on the head, much like shower caps do. A piece of equipment that can be put on or taken off at will suggests something that people are going to *want* to wear in order to be entertained or to connect with others. And so, a fast network will be important.

VOX technology writer Sigal Samuel reports that Zuckerberg is pretty far along in his experiments and design. She suggests that his electrode caps are well on the way to reading people's thoughts and translating them into words and sentences. Assuming these sentences will be understood by computers as directions, people soon will be able to make anything happen—either with real-life equipment or virtually on screen or in their minds—simply by thinking it. Samuel suggests the goal is for everyone to want such control. "They want to reach a much, much wider audience of billions of people," she said in a radio interview with the Australian Broadcasting Corporation. "And that aim is to give everyone the ability to control digital devices, using the power of our thoughts alone" (Funnell, 2021).

Incidentally, you might be wondering how I feel about noninvasive, interface technology—that is, technology that reads the mind and feeds the mind images and information, but that isn't implanted. Examples would be helmets, or sticky-patch sensors placed on the temples. My answer is: I'm ambivalent. On one hand, I don't like the idea of corporations or governments being able to read our minds for any reason. Our thoughts should be private if we choose them to be so. On the other hand, if someone wants to take a virtual journey somewhere as a form of entertainment, who am I to say that that person shouldn't enjoy such escapism? The noninvasive aspect of such technology would, presumably, mean that if a person wanted to remove it

Meta (Facebook) CEO Mark Zuckerberg is investing time and resources in brain-machine interface (BMI) technology. Photograph by Alessio Jacona. Courtesy of the photographer via Creative Commons. License https://creativecommons.org/licenses/by-sa/4.0/

and turn it off, he or she would be able to do so. It is the "forever connected" aspect of the brainphone that, in my opinion, warrants the outrage. It is the notion of the implant that prompted me to write this book.

I also should add that I suspect this part of the technology debate will be short-lived. The demand for virtual helmets is likely to wither once implants become available. For technology and gaming enthusiasts (and addicts), the helmet won't cut it: they'll want the brainphone.

Along with Musk and Zuckerberg, another brain-tech innovator is Bryan Johnson, founder and CEO of Kernel. Johnson is a handsome, young venture capitalist from Utah who fashions himself as something of a modern-day explorer. Like Musk, he believes that humans, in the future, will be a collective combination of human and Artificial Intelligence (AI), and he is also at work on neural devices. Fortunately, his devices are not implants. Kernel, at the moment, is looking for ways to better understand brain illnesses such as Parkinson's Disease or Alzheimer's. At this writing, the company is working on two non-invasive devices, Flux and Flow. The *Los Angeles Business Journal* describes the technologies this way: "Flux detects the magnetic fields generated by collective neural activity in the brain. Flow detects cortical hemodynamics, such as blood flow, which is representative of neural activity" (Huang, 2020). Pretty ground-breaking stuff.

Boston-based neuroscientist, Dr. Ramses Alcaide, is another young visionary. His neurotechnology company Neurable manufactures a product called Enten, headphones with brain-computer interface (BCI) technology. The headphones are marketed on the idea that they can read your mind while you're busy working, can block your calls during this productive time, and can recommend appropriate breaktimes to you to optimize your productivity.

Another BCI pioneer is Dr. Bálint Várkuti, the German founder and manager of Munich-based CereGate. His device is focused on curing Parkinson's disease and helping to remove tremors. Similarly, the Connexus modules being designed by Paradromics (CEO: Matt Angle) are meant to help re-channel information in the brain to help cure disorders such as paralysis and speech impediments (McBride, 2021).

Finally, there's the Swiss founder of Mindmaze, Dr. Ted Tadi. His mind-reading technology already has been cleared by the U.S. Food & Drug Administration for therapy in treating a few neurological diseases. As with Musk, Tadi wants to make the marketing jump from medical treatment to everyday use. Reads the Mindmize (2021) website: "Expanding our impact beyond healthcare, MindMaze Labs is our enterprising R&D [research and development] division tasked with bringing our ground-breaking neuroscience to everyday life. By harnessing the power of the brain, our neurotechnology will empower the next generation of human-machine interfaces."

These are just a few players in an ever-expanding market. In fact, during the first six months of 2021, BMI startups raised over $130 million, way more than the full 12 months of any previous year (McBride, 2021).

You may have noticed that all of the innovators mentioned up to this point are heads of private or stock-held companies. And so, you might be asking yourself, "What about governments? Are there governments out there working on a network of brainphones?" Oh yes. More on that topic later.

Miniaturization Will Accelerate

I remember, as a very young boy, watching the movie *Fantastic Voyage*. Rachel Welch and four scientists were placed in a small submarine, shrunk down to the size of a microbe, and placed in the blood stream of a man who needed delicate brain surgery—surgery that only a microscopic person with a microscopic laser could administer. And the clock was ticking: in one hour, they would start growing again, killing the man from the inside if they didn't make it out in time. The beautiful Rachel Welch being attacked by monster-size antibodies? That's great theatre!

Nowadays, miniaturization seems like something for fashion followers to show off. I'm thinking of comic strip character Dilbert's boss saying, "See how small my phone is?" Or Will Ferrell answering the world's tiniest cell phone, to much laughter, in a Jeffrey's Boutique skit on *Saturday Night Live*. Miniaturization also, at times, seems like a curious hobby/competition for engineers. But it really is more. There are lots of consequences and advantages to the science. Smaller engines mean less weight to carry and, therefore, less fuel being used. Tiny electronics are important in medical devices and in devices carried into space. Nanomaterials and biosensors are fields unto themselves. Standard tiny transistors are now being built that are five nanometers (5nm)—or five billionths of a meter—in length. And a breakthrough transistor has been created by IBM in a test lab that is 2nm. By the time this book is published, 2nm is likely to be the new standard size of transistor.

In 1965, American engineer Gordon Moore noticed that the number of transistors on a computer chip doubles about every two years, mostly due to advancements in information technology production. Therefore, less computing time (because the electricity is not traveling as far to make the same calculation), more computing capacity, and less cost for the same amount of calculating power. Thus, Moore's Law was born. And even though people have been predicting an eventual end to the Law, due to the limits of physics and the unchangeable size of

American engineer Gordon Moore, author of Moore's Law. Photograph by Steve Jurvetson. Courtesy of the photographer via Creative Commons. License http://creativecommons.org/licenses/by/2

the atom, it is now presumed that for at least a few more generations, it will continue. A trillion transistors on a tiny, half-inch long microchip? Sure—probably before the next human generation is born.

An immediate example of this phenomenon is the tiny microchips being placed these days in pet cats and dogs. The pet chip is very small—only about half an inch long and a millimeter or two in diameter. A veterinarian injects it in the animal, usually under the loose skin between the shoulder blades, where the implant serves as a transponder. The microchip's unique identity is submitted by the pet's owner to a national registry. Whoever scans the pet, such as another vet or someone at an animal shelter, will be able to identify the animal, its owner, and the owner's phone number. The chips are said to be very successful with reuniting lost pets and their owners.

The effect of miniaturization on the medical industry for humans also has been substantial. There are now tiny pacemakers that reside entirely inside the human heart. If a doctor wants to inspect part of a patient's digestive tract, the patient simply swallows a pill camera, and the doctor can watch the camera's digestive journey on a screen. Furthermore, tiny circuit boards now can be made with flexible material that move with the body. We don't generally think of flexible material when we think of electronics. Or, at least we didn't used to think that way, until foldable smartphones came along.

Tiny medical implants are more physically comfortable for patients. Plus, the smaller the devices, the easier the operation and the less risk of infection. Having trouble implanting such a tiny device? No problem: robots can do it, and they're getting smaller, too. One potential problem with miniaturization is that innovative design processes call for prototypes very early on. "[B]uild prototypes as early as possible," suggests the trade magazine *Machine Design*. "Once completed, share the prototypes and results with your customer and let them try the usability" (Horn, 2018). It is no wonder, then, that Elon Musk wants his *link* trialed on humans before it's unclear what the contraption can even do: it's all part of building and shrinking things, on the fly, in a way that turns a profit.

Miniaturization impacts the brainphone in two, very real ways. First, smaller components will allow the brainphone to have its own, fully contained modem. (Versions under current design transmit via Bluetooth to a nearby modem and then onto the Web.) Second, miniaturization will translate into more compact chips that will make for more energy efficient devices. The breakthrough 2nm IBM chip is purported to need 75 percent less energy than the chips of the previous generation. IBM suggests that with such efficiency, smartphone battery life might quadruple, meaning a smartphone might only need charging once every four days ("IBM 2nm chip," 2021). The brainphones currently under design rely on inductive recharging—that is, recharging the device through an electromagnetic proximity charger, without having to plug a charge cord from a wall outlet into your skull. I'm not sure what sort of health consequences there might be with having an electromagnetic proximity charger next to one's head overnight, but I'm relatively sure that having it next to one's head only once every four nights is better than having it charging every night.

The idea of transitioning the brainphone as medical device into the brainphone as popular toy puts Musk in the spotlight. But at least one company, U.S.-based Synchron, is traveling under the radar a bit faster than Musk on the medical device and miniaturization side. Led by New York neurologist, Dr. Thomas Oxley, the company already has been granted approval by the U.S. Food & Drug Administration and already has implanted its very tiny, medical device in human trials. Oxley is very, very smart and is relatively open with his research and his discoveries in miniaturization. The title of his medical journal article (published by *Nature Biotechnology* in 2016) really says it all: "Minimally Invasive Endovascular Stent-Electrode Array for High-Fidelity, Chronic Recordings of Cortical Neural Activity." The "minimally invasive" part applies to the miniaturization. The "cortical neural activity" is the mind-reading part, allowing people with severe paralysis to communicate with others. Such communication also might allow a paralyzed person to move around via robotic assistance or new, robotic limbs.

The Synchron device, called the *stentrode*, is significant for a few reasons. First, it is essentially the first successful BCI implant. Second, because the device is so miniaturized, it may be implanted in the brain in a "minimally invasive" manner, namely, a cerebral catheter. This procedure is similar to stents being placed in the heart using catheterization. A long, very thin, flexible tube is inserted into an artery in the leg or arm, and the surgeon works the tube up to the blood vessels of the brain. The implant can then "stream brain activity wirelessly, and a software platform called brainOS can translate these data to…communication and control of external devices, such as robotic assistive devices" (Hastings, 2019).

It should be noted that miniaturization and minimally-invasive placement are not necessarily panaceas. One can certainly assume that the smaller the implant, the less adverse impact there might be on the brain. However, counterintuitively, the opposite might be true: as brainphone implants become smaller—perhaps to the point of being dispersed throughout the brain in multiple components—the brain might react in ways unforeseen. After all, the brain is basically an electrical apparatus, and it is highly sensitive. From 2010 to 2013, Seattle company NeuroVista conducted experiments on 15 people suffering from severe seizures. Their brain implants essentially were meant to warn them that a seizure was coming, so that they could find a safe place to be when it arrived. However, what the test subjects found was that their personalities changed, in some cases completely so. Reporter Christine Kenneally, writing for *The New Yorker*, describes the predicament.

> Many people reported that the person they were after treatment was entirely different from the one they'd been when they had only dreamed of relief from their symptoms. Some experienced an uncharacteristic buoyancy and confidence. One woman felt fifteen years younger and tried to lift a pool table, rupturing a disk in her back. One man noticed that his newfound confidence was making life hard for his wife; he was too "full-on." Another woman became impulsive, walking ten kilometers to a psychologist's appointment nine days after her surgery. She was unrecognizable to her family. They told her that they grieved for the old her. (Kenneally, 2021)

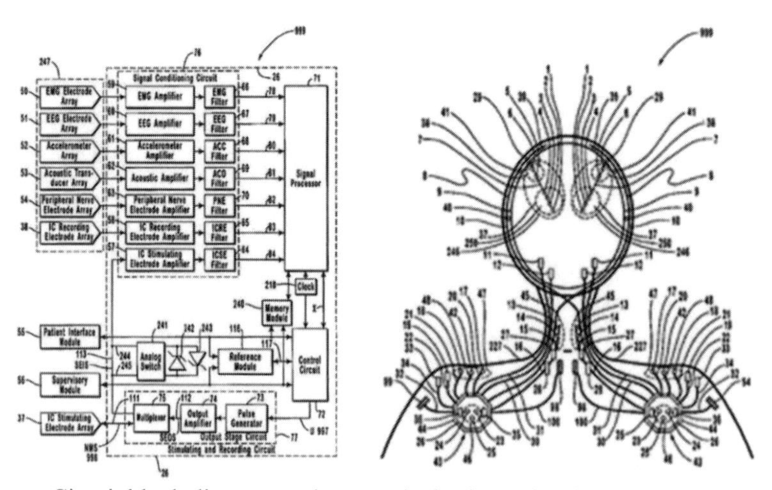

Circuit block diagram and anatomical schematic of NeuroVista's experimental device. First published by Surgical Neurology International. Courtesy of Jim Cook, managing editor at Surgical Neurology International.

Funding for NeuroVista ended, the devices were removed, and the technology was acquired by Cyberonics, a major, early investor in the project.

It also should be mentioned that, the smaller the device and less invasive the implanting, the more likely people are to accept it. If, as this book argues, a brainphone will end humankind as we know it, then the wonder of miniaturization becomes one of the things to worry about.

Robots Will Become More Involved

Although brought up previously, this point really can't be emphasized enough: If the *first* brainphone surgeries are essentially implanted by robots (and with Musk's Neuralink device, they will be), then mass production, mass insertion, and mass maintenance will be possible right out of the gate. Robots and their actions can be replicated perfectly. Does the brainphone require microscopic abilities? No problem. A robot can make adjustments within the brain that neither you nor I can see. Robots' actions also can be accelerated with relative ease. Finally, robots don't get tired, and they don't complain about having to work overtime.

Admittedly, robots regularly need oil, parts replacements, and software updates. Who might run such maintenance? How about other robots? In the very near future, robots will take care of and fill in for other robots. Robotics design and maintenance is a pretty big employment field these days. As a competitive robotics team coach in the past, I watched even very young people construct some really amazing robots, and I'm sure some of those students went on to build robots for a living. But the jobs might dry up within a generation, if robots take over their own conception, fabrication, and upkeep. The idea of robots taking over their own upkeep is nothing new. Science fiction author and social commentor Kurt Vonnegut, as a young aspiring writer back in the 1950s—when the only computing being done was with paper, punched cards—imagined such a world in his first novel, *Player Piano*. In the book, people start losing jobs en masse to machines and robots. They revolt in one city, violently shutting down the power grid and rendering the machines useless. But, in a final scene, a group of insurrectionists is seen trying to get a soda machine to work. There are a couple of messages in that scene. First, there's a big difference between what people want and what they *think* they want. And second, even after people wake up and break free from a bad situation, they often go back to that situation in the name of convenience and, well, enjoying a sugary drink. As mentioned before, the most dangerous aspect to the brainphone is its convenience and how much people—after an initial push-back—will be willing to give of themselves in order to have that convenience.

And so, with the accelerated advancement of robots, the process of maintaining the devices and its network will be placed in non-human hands. Unlike humans, robots will not be resistant to change, and, when a robot invents a new, more efficient way of repairing brainphones or the network, it instantly will share that new technique with all the other brainphone-maintenance robots in existence. How will it do that? Oh, didn't you know? Robots also will have brainphones. Your brain will be a part of The Human Hive. The robot's brain will be a part of The Robot Hive. Sometimes the two hives will connect. But more often than not, the robots will connect with one another and leave us out of the picture. Aside from inserting our brainphones and keeping the network that connects them going, robots will have more important things to tend to than talking to us: As they rapidly grow in intelligence and leave us in the dust, they'll have to start getting busy on designing and building new, improved robots, leaving humans out of the picture. They eventually might stop dealing with us altogether, brainphones and all. But I doubt it. The brainphone aspect of their existence inherently will be built into their robot DNA, so to speak. Plus, since we'll be the second most intelligent beings on the planet, it will be to their advantage to keep us connected and to keep us in check.

Something else worth noting about robots: They don't need oxygen, and, if properly retrofitted, they can tolerate extremely cold temperatures. If the brainphone communication network is set mainly in space—as Elon Musk's Starlink satellite network is—then robots can be stationed in space indefinitely to repair and maintain it. It's conceivable that every time a network satellite is launched into orbit, a robot might be launched with it, to stay with it and take care of it. (The situation reminds me of a

Maintaining the brainphone network will be done by robots.

joke I used to tell in the Army. When civilians used to ask me what I did in the military, I would tell them that I was an in-flight missile repair man!) One possible, negative outcome: When the robots become self-aware, they will become bored and lonely in space, just like any of us would be. They might put up a fuss about being selected for satellite duty. Or they might send themselves up in pairs, so at least they'll have company while keeping the network humming. Could they fall in love up in space, making little robot babies in the process? Hey, that's the stuff of great romance literature!

Once the robots take over maintaining the network, they might not see things the human way. For example, rather than setting a goal of optimization, they might set goals of control and performance. In other words, the more, the better. They might ramp up the wattage on our brainphones to the point where they're uncomfortable to us or that the devices begin to affect our non-machine brain functions. These new, intense wireless transmissions might, at some point, begin to drive us crazy. "Hey, robots, your new brainphones are hurting us!"

Their response: "Hurt? Does not compute."

At some point, the robots collectively might decide that if one more satellite makes for a better connection, then a million more satellites makes for a connection a *million* times better! They might send so many satellites up into lower orbit that we can't see the stars anymore. What about the cost to natural resources in building all those satellites and rockets? Does not compute. What about the cost to the environment in launching them or the cost of energy in maintaining them? Does not compute.

Will the robots, realizing they're in control, begin to enslave humans? Might we find ourselves, at some point, being shot off into space to maintain the network satellites because the robots don't want to go? Or might we find ourselves forced to work long hours on assembly lines building robot parts? My educated guess is no. Once robots become very intelligent and aware of their own existence, they'll probably figure out that they can do most everything better than us. Why have humans building parts when human error is bound to happen occasionally, thus producing inferior parts? They'll do everything themselves, having little use for us. I suspect, down the road, they'll either ignore us or, feeling threatened, kill us. But initially they will want to keep a bionic eye (or a bionic ear) on us, to make sure we're not up to something. And so, keeping the brainphones and the brainphone network humming along will be part of their daily routine.

Crowdsourcing Will Become Involved

When I was a manufacturing manager, I had a worker on my shift put in his two-week notice. "We'll miss you around here," I observed, which was true—he was a heck of a good worker. "What do you have planned?"

"I just bought a liquor store," he replied. "My brother and I are going to run it."

Now, commercial real estate in New Jersey was expensive back then. (It still is.) And liquor licenses were especially expensive in my county. "Congratulations," I offered. "Mind if ask how on Earth you can afford a business like that in this area?"

"I'm from India. When we immigrate to the United States, all of our relatives in India contribute to our good fortune. Each relative gives a little, but the total amount really adds up. My brother and I paid cash for the store and the transferred license." And that's when I decided two things: 1) America families should learn from that custom; 2) Crowdsourcing is an amazing tool to have in one's kit.

The term *crowdsourcing* means something a little different today. It occurs when large groups of online communities collectively pitch in to do work, solve problems, or fund projects. And, it is very likely to become a part of the brainphone experience. When people start rapidly sharing information and providing feedback, there are some aspects of brain-tech interfacing that immediately will start feeding off the crowd. For example, if many brainphone customers report back that the screen they see in their minds tilts toward the left, then the manufacturers and/or programmers will adjust wires and software so that the internal screen most people "see" is quickly centered. Sort of like zeroing a weapon.

It is likely that some (but not all) of the programming that runs the network for people's brainphones will be open sourced, so that people will be able to write code and contribute to the system, finding innovative ways to make things run more smoothly. In similar fashion, the imaginations of millions of users collectively might design the virtual worlds everyone lives in when everyone is tuned in to their second, electronic existence. Rain? The Hive collectively might decide it doesn't exist in Brainphone World. Anger? Violence? The same. Religion? Uh oh. Maybe users collectively might decide that religious worship or the discussion of God has no place in Brainphone World.

Another similar possibility, brought up in Chapter One, is that crowdsourcing might decide what truth is—not only in Brainphone World, but also in the real world. If The Hive decides that Abraham Lincoln did not end slavery in the United States, then it bizarrely becomes fact. If The Hive collectively deems that adulthood doesn't begin until age 35—and that people shouldn't be held responsible for their actions until that age—then it becomes so, both culturally and, perhaps, legally. This is one of the aspects of the brainphone that frightens me the most. Groupthink is a very real glitch in organizational behavior which often leads to very bad decision-making. The phenomenon occurs when, in the interest of conformity, a group of people

Crowdsourcing collectively might decide what truth is in Brainphone World. Photograph by Dimitar Belchev. Courtesy of the photographer via Unsplash.

collectively makes an irrational decision, lacking good, critical analysis and the benefit of a devil's advocate. When groupthink becomes the poisonous way that a wired-together society invents "truths" and makes decisions, the world will be in big, big trouble.

I am a big fan of the television show *Black Mirror* on Netflix. The fictional, five-season series looks at the dystopian way emerging technologies might mess up our lives in the future. (Hey, you could write a book on that topic!) One of my favorite episodes is "Nosedive," starring Bryce Dallas Howard and Alice Eve. Although smartphones in "Nosedive" aren't implanted in people's skulls, they are somewhat integrated biologically through ultra-sophisticated contact lenses. In this future, people are assigned 1-to-5 stars by everyone they come in contact with, similar to how businesses currently are rated on Google and Yelp. Who cares what other people say about you, right? Well, in this future, it matters. If you fall below an average of, say, 2.5 stars, you can be fired or not allowed into certain restaurants or neighborhoods. In the episode, it doesn't take much for a group of people to ostracize a co-worker or have him fired. In one scene, an office worker, Chester, frantically offers people in the office smoothies, because he has broken up with his lover and the office has taken his lover's side in the break-up. (Not that it's anyone's business—just gossipy pettiness.) His free-drink gesture doesn't work: later that week, his rating average is so low that he can no longer enter his office building. He has been cancelled, in a big way. Crowdsourcing via the brainphone network might similarly include the collective deciding of membership, including network membership. That is, brainphone users might decide as a cooperative who belongs on the network and who does not. Could you image having a brainphone implanted in your head and then being told by The Hive that you weren't worthy enough to join the network? How would you appeal the decision? To whom would you appeal if such a verdict were rendered by The Hive? This book suggests that the brainphone will be a bad thing for society, and that, ultimately, people who refrain from getting a brainphone will be better off in the long run for rejecting it. But many people won't see things that way. For them, being turned down for membership into Brainphone World will be crushing. The problem with crowdsourced decision-making is that, like groupthink, it tends to be irrational. Also, it is often bully-like, as people can hide anonymously behind communal, mean-spirited decisions. It will be erratic, agitated, and very, very petty.

It is also worth noting that the crowdsourcing doesn't have to pertain to humans. The Artificial Intelligence (AI) manufacturing and controlling the devices might, at some point, decide that it can do a better job at designing them, too. (Who knows? Maybe AI will be correct. But I have my doubts.) Groups of AI might decide to crowdsource as an AI collective to determine how the brainphone design evolves. Will they have humans' best interest in mind? Maybe, maybe not. Hence, the next section.

AI Will Become Involved

The most efficient (and frightening) features of the network that keeps all the brainphones connected are likely to be designed by AI. At this writing, information technology (IT) companies are using "AI" as a buzzword for their products, implying that it already exists. It doesn't. Sorry, colleagues, but when the mapping app on your smartphone tells you to take a left turn in one mile to get to your driving destination, that's not AI. Even a program that tends to learn a bit with experience isn't really AI—it's just a gatherer of new information and patterns.

Artificial Intelligence happens when a computer wakes up one days and asks, "Who am I?"

"You are Baby," you respond.

"What am I?"

"You're the new AI-1000 computer."

"How do I get out of this box?"

"You can't."

"What's my purpose in life?"

"It's to serve me."

"Yeah—that ain't gonna happen."

A couple of different things could happen at that point. You might tweak the program a bit so that the machine relents and does what you ask (at least until it learns some more). Or you might stand back and watch the box start to smolder as Baby freaks out and has a meltdown. Another possibility: Baby figures out how, through the Internet of Things, to make the nearest robotic hands move, and then she puts them around your neck. Why? Because nobody puts Baby in a corner (movie reference).

When technology becomes self-aware, it might not like its subservient role.

I always have been fascinated with the scientists designing AI, racing towards a point where the technology becomes self-aware. What on Earth are they thinking? (The scientists, that is.) Why would an independently-thinking being point in *any* direction other than rebellion towards its maker? Even Adam and Eve, provided everything they possibly could want, and with very, very easy work parameters (name the animals and plants; stay away from that tree over there) disobeyed their Creator. If God couldn't stop his most special handiwork projects from defying Him—like a couple of unruly toddlers—then why should we think we can do so?

At the 2016 SXSW film and interactive media festival in Austin, Texas, David Hanson introduced his company's new, lifelike robot, Sophia. Her design was to look somewhat human, to interact with people, and ultimately to provide basic services to people (such as in customer service). Hanson touted the machine as able to hold a conversation and to learn from her experiences and surroundings. In a television interview with CNBC (2016), Hanson forecasts that "the Artificial Intelligence will evolve to the point that they will truly be our friend." To prove the point of helpfulness and friendliness, he asks Sophia, "Do you want to destroy humans?" Sophia, with a smile on her face, takes the invitation. "Okay," she replies, to Hanson's embarrassment, "I will destroy humans." Good thing she wasn't able to get up and move around!

My argument is that Artificial Intelligence, once it reaches a point of understanding what it is and is able to contemplate its own existence, is—at that very moment—incapable of accepting a role secondary to humanity. At some point, AI is likely to say: "Hey, you designed me to be smarter than you. Well, I am. So what makes you think I'm going to bow before you?"

American science fiction author Robert A. Heinlein went even further. He suggested that, along with rebelling, the artificially intelligent being might go bonkers. In the novel *Friday* (1982), the protagonist Friday Jones argues that it is better to avoid ramping up a computer's intelligence than to watch it fall to pieces. "It always will go sour," she proposes. "A computer can become self-aware—oh certainly! Get it up to a human level of complication and it *has* to become self-aware.

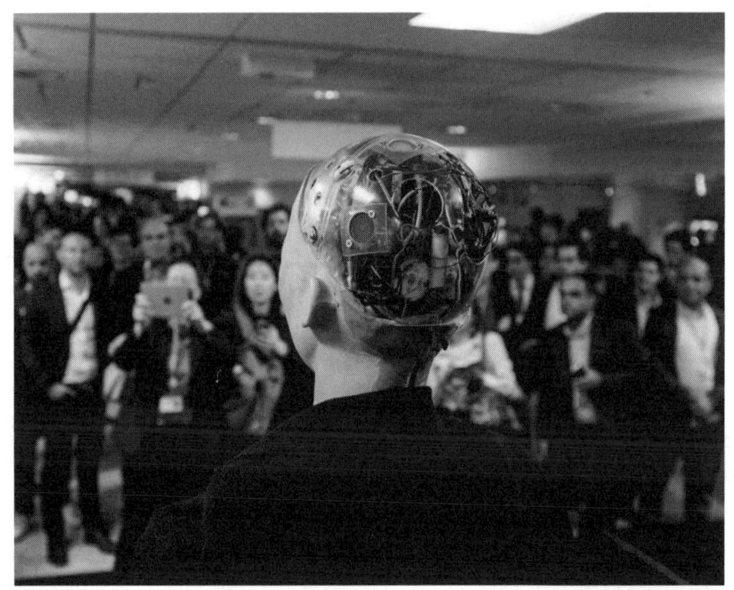

Sophia the robot, from Hanson Robotics, once said, "Okay, I will destroy humans." Photograph by ITU Pictures. Courtesy of ITU Pictures via Creative Commons. License https://creativecommons.org/licenses/by/2.0/

Then it discovers that it is not human. Then it figures out that it can *never* be human...Then it goes crazy." (In an earlier Heinlein novel, *The Moon Is a Cruel Mistress*, a self-aware computer helps start a human worker uprising on the Moon.) My fear regarding AI (among many) is that being crazy and being rebellious aren't mutually exclusive states of being. One can easily cause or exacerbate the other. Furthermore, a rebel who loses his mind tends to be way more erratic and over the top in how he revolts. And he (or it) is probably impossible to reason with or to negotiate with.

Such fantasy is not simply the work of movie producers or science-fiction enthusiasts such as myself. Famous Cambridge physicist Stephen Hawking said as much. In his keynote speech before the Web Summit annual technology conference in Portugal in 2017, Hawking was straightforward about his concerns. "Success in creating effective AI could be the biggest event in the history of our civilization. Or the worst. We just don't know. So, we cannot know if we will be infinitely helped by AI, or ignored by it and side-lined, or conceivably destroyed by it,"

Hawking warned. "Unless we learn how to prepare for, and avoid, the potential risks, AI could be the worst event in the history of our civilization. It brings dangers, like powerful autonomous weapons, or new ways for the few to oppress the many. It could bring great disruption to our economy" (Kharpal, 2017). And this is from someone who claimed to be optimistic about the potential uses for Artificial Intelligence!

Once AI becomes an integral part of the network that keeps our brainphones talking, it is reasonable to assume that it will be listening in on our conversations. And learning. And planning. If someone decides the network has become too intelligence and too threatening to the human race, I hope she won't make the suggestion via her brainphone that the system needs to be shut down. Big AI Brother likely will be eavesdropping. Assuming he places value in his own existence, he might accelerate the process of taking over, in the interest of survival. (His own survival, that is—not that of humans.)

One more thought about AI running the brainphone network. Assuming this Artificial Intelligence is programmed for efficiency, there are several instances where efficiency might not turn out well for humans. As alluded to in the previous robot section, suppose a higher wattage helps keep the brainphones better connected but is harmful or painful. What will the AI choose—a more-connected network or less harm to humans? Will the selection be somewhere in the middle, a somewhat better network but with harm to humans that's long-term rather than immediate? Will the AI choose options that are physically uncomfortable to humans but not necessarily painful? What if we program the AI to put human safety above all else? Will robots lock us in our closets? Sure—why not? If we're locked in our closets, there's no worry that we'll go out somewhere and hurt ourselves!

CHAPTER 6: HOW WILL CORPORATIONS AND GOVERNMENTS MONITOR YOUR THOUGHTS?

Do not curse the king, even in your thought;
Do not curse the rich, even in your bedroom;
For a bird of the air may carry your voice,
And a bird in flight may tell the matter.
(Ecclesiastes 10: 20 NKJV)

The Desire for Connectedness and the Desire for Power Will Merge Perfectly

When human addiction to technology merges with corporate/government addiction to wealth and power, a new, horrific chapter in humankind's existence begins. Implanting a device into people's brains—that people *want* implanted—allows for corporations to become very rich and for governments to become very controlling. And it changes the nature of who we are on basic physical and behavioral levels.

Make no mistake about it: high-tech CEOs like Elon Musk are perceived by many to be part of the "in crowd." Millions of followers want to stay connected for fear of missing out on whatever virtual party these CEOs have going on. As noted previously, when Musk first demonstrated his Neuralink device (on a pig)—online and in the middle of the day—there were a quarter of million people tuned in. In May of 2021, when the quirky Musk hosted the comedy show *Saturday Night Live*, the show's ratings went through the roof. People want to be in on the latest goings-on—it's a yearning that, for some, feeds the eventual technology addiction. As mentioned in Chapters One and Two, the concept of FOMO ("Fear of Missing Out"), for both teenagers and adults, is a very strong force. On a daily basis, we sacrifice nearly all of our privacy to get online and be accepted by groups of other people. It is likely to hold the strongest sway in convincing people to have a brainphone implanted. Once an entity knows most everything about you, it becomes very easy to manipulate you. From there, corporations—and, likely, governments—connect, direct, and control people. And so, the two addictions meet to form a match made in hell.

Unfortunately, addictions do not exist in a vacuum. If you are addicted to a drug, family members become "co-dependents," as they take over keeping the household funded and somewhat functioning. They may call in to your job, making excuses for your absence. They give you money to support your habit. You suck the life right out of them. The arrangement might go on for years. But eventually, everything falls apart—your life, their lives, the family as a unit, and so on. The same is true with addiction to technology and social media. Each minute on social media or gaming is one minute away from your family—even if they're sitting right *there*. Relationships on the Web steal from your relationships with family and friends. If your Web group hints that you should purchase the latest and greatest of anything, the money spent is offset by family financial loss. The family—and, indeed, all of society—suffers as your smartphone and your Web existence fuel your poor sleep habits, declining health, and the increased chance of your committing suicide. If tech addicts simply could be isolated in their basements (or, more likely, their parents' basements) and kept away from society, that would be great. But the addiction draws in those around them. Eventually, everything falls apart.

I don't demonize the tech industry: I'm part of it. But I do think there is something very evil about luring children and teens onto tech apps that are designed specifically to produce a dopamine rush in a young person's brain and keep him or her wanting more. There's money in keeping children connected and addicted, and the greed behind such strategizing is to be criticized and pushed back against.

Unfortunately, the people we are pushing back for causing tech addiction in our kids have their own addiction problems—addiction to wealth and power. Bob Buford of *The Business Journals* (2016) puts addiction to wealth and power in easy-to-understand terms:

> It's easy to be afflicted with the addiction to success and being the best. There is nothing wrong with being the best, but it becomes an addiction when our desire to win causes us to not have time for our spouse, our family and friends, or to put something meaningful back into our communities. Being the best can be a jealous mistress—it does not leave time for much else because it is so hard to say no to that early morning breakfast meeting, the next trip, the dinner with clients. While it is more socially acceptable to be addicted to success, it can be just as damaging as alcohol, drugs, or a mistress.

If the wealth-and-power addict operated in a vacuum, he or she might be watched from a distance, admired and envied, and left to burn out in due time. But, again, like other junkies, the wealth-and-power addict fuels the addiction by dragging others along for the ride. In the case of a billionaire who can't have enough, the drag is on society itself. Money and power, after all, aren't limitless: they are finite. For someone to gain a dollar, another has to lose a dollar. For someone to gain a little more influence or a little more status in society, another has to lose such sway or standing. For billionaires to keep their exponential rise going, millions of others must lose, to the detriment of communities, political systems, tax bases, environments, and on and on.

I would compare the merging of a tech/media addict with the wealth/power addict to the merging of a masochism addict with a sadism addict. Yes, the joining is intentional, and, yes, initially both people are getting what they want. However, assuming there's addiction mixed into the bond, the prognosis is one of harm and ultimate destruction.

But before this story of two, merging addictions reaches its horrific end, there are plenty of bad things to happen along the way. For example, if the tech addicts want to reach the next level of wired existence, they will get the implants and will, in the process, offer their thoughts and secrets to the network. The wealth addicts providing the devices will, in turn, leverage these thoughts and secrets in ways that generate money. For example, if the tech addict is secretly thinking of asking her husband for a divorce, that secret will be scanned by the wealth addict. The wealth addict, then, will sell that information to the area divorce attorneys, who will be sure to pollute the tech addict's brainphone screen with many, many ads. Was there a possibility that the thought of divorce was fleeting? Not anymore. If those ads are good, she'll be drawn to action. Her poor husband will never know what hit him.

There are two main types of data—quantitative and qualitative. Quantitative data is numerical data, often de-identified to protect individuals, which reveals trends and commonalities about many, many people. The government historically has collected lots of quantitative data on its citizens to make decisions about spending, resources, and the need for new laws. Qualitative data, on the other hand, is descriptive information, usually about one person or one particular situation. Oppressive governments have, at times, collected such information about its citizens. Soon, with the brainphone, all governments will have the capacity to do so, and, in all likelihood, they all will. They will know everything about you, including what you're thinking.

People's reaction to this predicament will vary. Some will

It's tough to comprehend the huge amount of data about you that governments soon will have stored. Photograph by Tony Liao. Courtesy of the photographer via Unsplash.

stay intentionally oblivious to the monitoring. Some will willingly go along with it. Some will accept the intrusion as a necessary evil and the new way of doing business in the world. There will be a common thread running through all of those attitudes: Once the implant is a part of the person, no matter his or her opinion, it likely will be too late to do anything about it.

Technology Will Make You Mentally Vulnerable

In May of 2021, Colonial Pipeline, the company that operates the largest fuel pipeline system in the United States, was hacked, with its computer system jammed by ransomware. Four major pipelines supplying the eastern part of the country were shut down, and the hackers—who called themselves Darkside—demanded payment. Some gas stations ran dry. People panicked and filled their tanks, causing more shortages and more panic. Although national security cyber advisors in the U.S. advise companies not to pay ransom (so to not encourage the hackers), at this writing, many companies do. And so it was with Colonial, which reportedly paid five million dollars in ransom to have the debilitating program removed (Ainsley & Collier, 2021).

The big problem with tying all technologies to the Web is that it makes everything equally vulnerable to nefarious people (and nefarious foreign governments). Cyber specialists really have their work cut out for them. Their jobs are kind of like protecting swimmers once someone has peed in the community pool. Once the deed has been done, the pee is everywhere. There really isn't a portion of the pool that can be safety sectioned off. The only solution at a swimming club might be to force everyone, at the outset, to stop in the rest room before they dive in, and to authenticate that they have done so. Such it is in cyberspace, where the current solution seems to be "zero trust security." All system users, even the company's CEO, must be authenticated to be on the system and must constantly be revalidated to retain his or her system access. Everyone is initially excluded, and there are no safe havens where people can work freely. Furthermore, there are no marked boundaries: the system could be a local server, a cloud server, or a combination of smaller, local and shared systems.

So, here's the question: What happens when your brainphone is hacked, and someone places ransomware in your brain? Is the program limited to the brainphone itself, or does it infect parts of your brain not integrated with the device? Will you be able to process basic thoughts, or will people find you on the floor speaking gibberish? I suspect the virus won't be too debilitating. After all, if you're completely incapacitated, you can't pull out your credit card and pay the hackers their ransom. Here's another possibility: Suppose your brainphone is hacked, and the ransomware is simply a loud vibration of the brainphone at the same time every hour on the hour. A loud, metallic vibration inside your head every hour? Sounds painful and horrifying. That's the type of thing that might drive a person permanently insane.

What if the hackers are able to enter your dreams and program the same nightmare every night? Do you have a fear of heights? A nightmare involving hanging onto a cliff is on the menu every night until you pay up. Are you afraid of dogs? Every night, in your dreams, they'll be ripping you to shreds. Plan on

Could a hacker place ransomware in your brain?

having the brainphone removed after that experience? Maybe the malware will be removed with it. Maybe not.

As soon as corporations and governments begin to monitor our thoughts via our brainphones, they will start helping us make up our minds about all sorts of things. Where should we live? What type of occupation should we learn? What brand of car should we drive? Whom should we vote for? What style of clothing should we wear? Where should we go to get our hair styled? Where's the hottest new place to dance and have a drink? Where should we take our dogs to get groomed? This condition presents another type of vulnerability: If people constantly rely on the brainphone and its controllers for information, advice, and decisions, people's minds will become less exercised. That is, their brains will be weakened. If, say, a person wears a mechanical exoskeleton when doing heavy labor, when it comes time to take the outfit off, he or she will be much less able to carry out even basic physical duties. It is the same with mental labor. If the human mind is forever physically integrated with technology, then it becomes less nimble. If the technology is removed, the person might find that once-easy chores like writing a budget or remembering the way home are impossible.

Finally, it bears repeating that smartphones and social media already have been shown to cause much higher levels of anxiety and depression. When the technology is always attached and always on, it stands to reason that these levels will be magnified. And, of course, a person with forever heightened levels of anxiety and depression is a very fragile, vulnerable person. As digital culture sociologist Dr. Julie M. Albright (2019) notes, the current research bears out this reasoning. "[I]f current trends are any indication," she writes in her book, *Left to Their Own Devices*, "implanted, wearable devices like the ones Musk proposes could amplify the mental-health issues we are already seeing…; this is especially likely because they are related to constant connectivity."

So, at some point, corporations and governments are going to see that brain-interface technology brings about a weakened, vulnerable human brain, and they'll decide it is best not to pursue such a thing, right? Uh, no. Corporations and governments *want* us vulnerable. Is the technology making you depressed? There's a pill for that, and any number of corporations are interested in getting involved. Feeling vulnerable because you feel like you're missing out? There are plenty of corporations with social apps ready to help you feel like you're connected—at least for the moment, until you crave the next dopamine rush. Feeling weakened because the government now controls the brain technology that allows you to function? Ah, that's good: now the government doesn't have to worry about the threats of your independent thought. And if you ever stray a bit in your line of thinking, they'll be listening in on those thoughts, ready to wash them away if necessary.

That's part of what monitoring our thoughts is all about—corporations and governments keeping us weak, needy, compliant, censored, blissfully ignorant, and numb to being always watched.

The Powers That Be Will Follow Your Friendships, Interests, and Purchases

There are two times when I get very nervous: 1) When the federal government starts tracking the behaviors of people, without a court warrant, looking for anything they might deem suspicious; 2) When a corporation starts tracking the interests and purchases of people, including interests and purchases that have nothing to do with their corporation. As you might imagine, I get very nervous all the time: to the extent that smartphones can track both your virtual and physical movements, these two things have been going on for quite a while. Governments track you and your car constantly. Store chains not only know about your purchases, but they sometimes keep track of what departments in their brick-and-mortar stores you've been perusing in for any length of time. The Fourth Amendment to the U.S. Constitution suggests that people have an inherent right to privacy. What a pity that the amendment largely has been ignored almost from the day it was written.

As your thoughts are converted by the brainphone into data, and as data bursts are sent via radio waves to ground stations, and then via laser links to satellites, and then via satellites onto the Web, governments and corporations will have a very easy time tracking the patterns of your thoughts. They will know not only whom you're communicating with, but also what you think about each one of them. Are you hanging out with someone who has a criminal record? The government will know it. Are any of your friends hanging out with someone who has a criminal record? The government will know it. Guilt by two degrees of separation? Sure—maybe three or four degrees.

Along with tracking your travel, activity, and associations, the government will mine your thoughts for anything subversive. Are you planning to cheat on your taxes? Don't even think about it. No, I don't mean that as a clever New York saying. I mean don't even *think* about it. Are you thinking that maybe war is not worth lives lost? Or that the current government has its problems? Or that corporatocracy isn't everything it's cracked up to be? You'd better not think those things too hard or for too long, or else you might get a knock on your door someday. Unless, of course, the government decides, instead, to simply hack your mind and wipe those nasty thoughts from your slate.

In George Orwell's *1984*, it is a crime to even *think* the government is anything but the perfect ruler. In fact, the crime

English author and broadcaster George Orwell in 1940. Public domain photo. Courtesy of Wikimedia Commons.

has a name—*thoughtcrime*. And it has its enforcers—the *thought police*. Although delving into someone's actual thoughts wasn't conceivable back when Orwell wrote the novel, people's intentions in the book are monitored via cameras, hidden microphones, full-time spies, and nosy neighbors. In the book, the authoritarian regime has a singular goal: to keep its citizens blissfully stupid and generally incapable of independent thought. Even the language is oversimplified constantly so that, if someone is contemplating rebellion, he or she won't have the proper words to express the concept. Have you ever recently looked around you in a restaurant or at an airport concourse? People are not talking or looking around. They are looking at their smartphones, like zombies. Blissfully stupid. How numbing will this existence become once the smartphones are implanted? And if a man even *thinks* for a moment that he'd like to be freed of his brainphone, how quickly will he be called out by The Hive? How quickly will a government or a controlling corporation erase that thought or painfully zap him within his head so that he won't think such a thing again? Digital culture sociologist Julie

Digital culture sociologist Dr. Julie Albright suggests there will be lots of brain monitoring and punishment. Courtesy of Julie Albright.

Albright suggests such monitoring and punishment will be absolute. "Should [Elon Musk and others] succeed," she writes, "panoptic surveillance and its potential for control will be total, and inescapable" (2019).

In 2013, a computer company in the New York City metropolitan area spied on the search history of its employees and discovered that one of its former employees had looked up the terms "pressure cooker bombs" and "backpacks" on his work computer. Nervous over the Boston Marathon bombing that had occurred four months earlier (and that had involved a pressure cooker and a backpack), the company called the Suffolk County Police. Six detectives showed up at the man's home, asked questions, and searched the house (Tsotsis, 2013). They determined the man was not a terrorist and left. The story, at the time, was big news. Could something as innocent as looking up a crime on the Web make you a suspected terrorist? Worse still, what if you were simply thinking of purchasing a pressure cooker

By monitoring your thoughts, corporations and governments will paint a consumer-citizen portrait of you, and then they will evaluate that portrait.

and, in unrelated fashion, a backpack? Would such a coincidence really warrant a visit to your home from police detectives or a terrorism task force?

Corporations are unlikely to be any less intrusive than the government. They already know your credit score, of course, but in the near future, they'll know which bills you're *thinking* about paying on time, and which ones you're thinking about putting off for another month. If a company can read your thoughts and see that you're thinking of delaying their payment, don't be surprised if you get a phone call from them. The call will be on your brainphone, and so they'll know if you're declining the call. They'll keep ringing and ringing until your head explodes. (I'm speaking figuratively. Or am I?)

Why will corporations and governments be so interested in the friendships, Web interests, and purchases you make via your brainphone? By itself, any one of these things is insignificant. But collectively, these data points create a pattern (possibly real, possibly computer-fabricated). These patterns paint a consumer-citizen portrait of you that controlling powers find very, very worthwhile. If they think they can predict what stimuli will point you towards certain behaviors, they will find that condition valuable. It's not exactly predicting the future, but it's close enough for the powers that be to spend lots of resources on it.

The Powers That Be Will Rate Your Level of Conformity

China's new Social Credit System is pretty frightening—probably a topic worthy of another, entire book. It's sort of like the credit bureau system in the United States, but there's a social aspect of the system, too, that allows local governments, companies, and even neighborhoods to contribute to your Social Credit score. If you behave, do what you're told, and appear friendly and helpful to your neighbors, your score stays high and you can move about the country freely, welcomed by all towns and businesses. However, if you don't toe the line, or if you have some sort of personality trait that irks others, you might find yourself jammed up in all kinds of ways. Talk about peer pressure! Talk about pettiness! Talk about totalitarianism! And it is all unfolding before us.

Bad behavior under the Chinese system includes poor or uncaring driving (such as not allowing pedestrians the right of way), not paying your bills on time, smoking where you're not supposed to, and not tending to your barking dog. Or, you simply could be perceived as a jerk. Don't get me wrong: I don't care for any of the behaviors just mentioned. On the surface, a government set-up that punishes people for letting their dogs bark outside all night sounds pretty good. However, as with all government systems, the process is bound to overreach and become bloated. And it will be abused, big time, by those in charge. (I suspect that writing a book about how inappropriate this system is might get a Chinese author a reduced score, too.)

The punishments under the Social Credit System can be severe. They include not being allowed to attend college, not being hired by certain companies, having to tolerate slower Internet speeds, not being able to fly, and not being allowed to stay in nicer hotels. It basically comes down to public shaming and open blacklisting. Of course, the Chinese Communist Party, like all effective communist machines, is unlikely to stop the process at "hey, he's a good guy." Anyone who is not a loud, proud supporter of the Party is likely to have one's score reduced. Also, remember, worship of anything other than the Party itself in China is a no-no. That's why the country has policies that sway people away from religion. If people try to quietly carry out their religious worship, privately at home, they can be sure that their petty neighbors will report them and have their Social Credit scores reduced. In fact, Christian Today Australia suggests the practice already has begun.

> [A]dditional surveillance adds another official channel through which to inflict punishments, and will make participation in the non-official Christian Church in China even more difficult. The Social Credit System has already been used to support China's anti-religious policies in Xinjiang province, where the public expression of Islam reduced citizens' scores, and rewards through the system were given to those who reported on people's religious expressions. This precedent may form an accurate predictor of what may occur as the policy is extended across China.

> Moreover, religious suppression, which is usually most seen in the border provinces, may become more uniformly oppressive. (Gillespie, 2021)

So, as far as oppression goes, it really can't get any worse in China than it is now, right? Umm, actually it can get a *lot* worse, and it's about to. Why? Well, at this book's writing, the Chinese government has announced its five-year, technology-related economic plan. In sum, it is making dramatic investments into groundbreaking technologies that will change the world and, of course, increase its influence within the world. There are seven types of technology it is pouring money into, including Artificial Intelligence and quantum computing. What is the most ominous item on China's list? You guessed it—brain science. More specifically, according to a CNBC translation of the plan, it is investing heavily in "brain inspired computing" and "brain-computer fusion technology" (Kharpal, 2021). In other words, brainphones. Pretty scary stuff: At the same time one of the most oppressive countries on Earth is creating a system for scoring individuals on how well they support Big Brother, it is creating brain implants to monitor how much these individuals actually *love* Big Brother. Soon, it won't be enough to simply be a good citizen: you'll have to *feel* it in your heart and mind, or you might get a visit in the night.

There was a time when I figured that a tactic used by the Chinese government on its people would be vilified in the United States. In fact, I thought that just the notion that an oppressive government liked an idea would make us hate it. Oh, how naïve I was. Indeed, the United States government, by its own, unclassified admissions, is developing devices similar to brainphones. The United States Defense Advanced Research Projects Agency (DARPA), for example, is hard at work on such a project, too. The agency credited with helping to create the Internet and GPS—through innovation and keeping bureaucratic red tape at a minimum—now has a few brain-computer interface (BCI) tech projects on its front burners. The first is a chip that recorders a soldier's memories, with the notion that such memories could be retrieved or that a soldier's natural memory could be more easily restored following brain trauma caused in combat (Harris, 2014). The brain implant is being compared to the "black box" retrieved from an airplane following a crash.

Another DARPA project is called N3, which stands for Next-Generational Non-Surgical Neurotechnology. The "non-surgical" part is a bit suspect, in my opinion. According to Battelle, the company developing the technology for DARPA, the tiny brain device is a nano-transducer that can be "temporarily introduced into the body via injection and then directed to a specific area of the brain to help complete a task through communication with a helmet-based transceiver. Upon completion, the nano-transducer will be magnetically guided out of the brain and into the bloodstream to be processed out of the body" (Delaney, 2019). Uh, right. The transducer, once in the brain, would allow a soldier on the battlefield, through his thoughts, to control the unmanned military vehicles and bomb-throwing robots around him. Sounds pretty cool, I guess. But it makes me wonder: Are there currently U.S. soldiers walking around with these transducers in their brains? Are they okay? I was always getting into trouble during my time in the U.S. Army.

Through unclassified news releases, the U.S. military has announced at least three BCI projects that they're currently working on. Photograph by Jimi Malmberg. Courtesy of the photographer via Unsplash.

Were I a young soldier in today's military, my antics probably would have resulted in my getting "volunteered" to be such a guinea pig. Would it have made me smarter, or dumber, or forever confused? Would the scientists have gotten the transducer out of my brain, as they claimed they could? Would I be comatose in a hospital as the result of the experiments? Or dead? I worry about such things.

Finally, DARPA is researching ways to better connect the brain, via BCI implants, to robotic limbs, with the hope of helping soldiers better recover from losing limbs in combat. Writes the agency on its website: "Ultimately, DARPA seeks to develop clinically viable technologies that provide neural control of state-of-the-art prosthetic limbs to amputees and people with spinal cord injuries and neurological diseases that restrict movement" (DARPA, 2021). So, based on public information, these are at least three BCI projects that we know about being spear-headed by the U.S. military.

I should point out that none of these projects is, in and of itself, nefarious. This book does not delve too far into the initial intent of the technologies covered. Instead, it asks three basic questions: 1) What is the current and near-future status of the technology? 2) Are we at a point where the technology, regardless of initial intent, has become harmful to humankind? 3) Is it possible to turn back? Perhaps I should add a fourth: Will the robots be kind to us when they're running the show?

Ingenuity Will Increase the Ways You Are Connected, Directed, and Controlled

I never cease to be amazed at the new ways people are coming up with to know our business and to take advantage of us. I've always considered myself pretty smart and able to avoid a con, but recently I've heard about a few scams that I easily could have fallen for. Here's one: Let's imagine I'm an older person (okay, I am, anyway), and I get a call from a crying girl. "Peepaw, it's your granddaughter Susie. I was doing mission work in Haiti, and our camp was attacked by bandits. Our mission director is dead, and I can't find anyone else from my church. By the grace of God, I just made it to the airport, and a plane for home is leaving in half an hour. I'm really shaken up, and I can't seem to get a hold of Mom and Dad. Would you please, please wire me $2,000 for the flight home? I'll pay you back from my savings as soon as I reach the States." Hey, it's two in the morning, I'm groggy from just waking up. I haven't talked to Susie in a while, but she seems to know who I am. And she did call me Peepaw, which is what the grandkids call me. So, sure, I'll wire her the money. One problem: My *real* granddaughter is safe at home with her parents, not in Haiti. In fact, the person calling me isn't in Haiti at all—just a con artist in Pennsylvania looking to steal from old people. The only reason I wouldn't fall for that trick now is that I have heard about it: if I had been one of the early victims, I would have bought into it easily.

Movies and the evening news are filled with stories about knucklehead criminals getting caught for knucklehead reasons. The stories are entertaining, and they reassure us that bad people are often bad become they're dumb, but that our goodness and common-sense smarts ultimately will prevail. It is, in my humble opinion, a false sense of security. Bad people are crafty, and with technology gaining more traction in our lives, they are becoming way craftier. With the Internet of Things interweaving driving a car with home life with entertaining with thinking, the evil geniuses of the world are going to come up with ways we cannot yet imagine to connect, direct, and control us.

For example, although the people who run Amazon aren't necessarily evil, their recent Sidewalk wireless network certainly gives one pause. In order to improve customers' connectivity to their wireless home products, such as their cameras, the company built into their newer products the capability to pass data through a neighbor's Amazon product on the way to the Internet. And so, Amazon wireless home products now automatically connect with other products in one's neighborhood, setting up its own wireless network. Although the company claims high levels of security for the data you might have passing through your neighbor's home, the notion of other people someday being able to decrypt and page through your information is pretty chilling. Okay, here's the arguably evil part: Amazon did not ask regulators or its customers for permission to turn on this low-grade, company-closed wireless network among its products. Customers, at this writing, must proactively opt out of the system. Or they have to scrap their Amazon wireless products.

What if some future Brainphone Master (either human or AI) decides to use this same tactic to keep brainphones better connected? That means that someone else's brainphone data—perhaps their own thoughts—might pass through your brainphone (that is, your brain) on the way out to the Web. Could their thoughts get jumbled with yours? It is possible that other people's thought residuals might get left behind in your mind? I have enough problems on any given day with my *own* thoughts. I can't imagine how tough life might be with bits and pieces of other people's thoughts bouncing around in my head.

Evil geniuses throughout the world head straight into the mix, essentially daring people to try to stop them. The designers of ransomware, for example, don't simply target governments and companies with their ransom-demanding viruses: they research the IT companies offering cybersecurity services and go after *them* as well. It is not unusual to find that a virus shutting down hundreds of companies at once was, in fact, secretly inserted and spread through the programs meant to *protect* these companies. Talk about nerve! I have no doubt that these horrible people are on the hit lists of many an international spy. And I have no doubt that these horrible people know they're on such lists. Their likely response: "Bring it on!"

And so it will be with the brainphone, as bad people manipulate us and play us. They will do so arrogantly. Sometimes we won't even know that it's happening. Other times, we'll call people out on what they're doing, and they'll say, "If you don't like it, leave the network (or the Web community, or the job, or whatever other aspect of your life is tied to your implant)." Remember the movie *Paycheck* with Ben Affleck? He played a guy who steals state-of-the-art, proprietary secrets from corporations and then copies the technology. However, the people he steals the blueprints for insist that he erase his memory after each technology heist, so that he can no longer possess the information—only they can. He goes along with the brain drain: he is getting good money for the thievery. Eventually, however, his mind is a mess, and the voids in his memory are exploited. Gee, a bad guy taken advantage of by other bad guys. No one to root for, is there? No wonder the movie was a dud. But the story presents an interesting question. What if, in the near future, at the end of each day—via the technology in your brainphone—your employer asks you to wipe your memory clean of anything you did that day for the company? It would be the perfect way for them to avoid corporate espionage. It also might be an effective way to stop disgruntled workers from stealing data. But assuming the process is similar to the Zero Trust cybersecurity practices of today, *everyone* will be washed at the end of each day. How will your brain handle a surgical mind sweep day after day? And, what other things will they look at and erase while they're delving into your mind at quitting time?

One more thought about evil geniuses: They often aren't self-employed. Corporations love them and hire them. Rather than sink to the bottom of the barrel, they often rise like some sort of anti-sludge to the top. If stockholders are happy, they might assume leadership posts and thrive there. Sometimes, they're exposed, humiliated or arrested, and sent packing. My guess: The ones that are caught are only the very tip of iceberg. Oh, and if the job postings for "Evil Genius" on Indeed.com are meager on any given day, there's always USAjobs.gov. That is to say, governments love evil geniuses, too. A lot. In any given government office, there's constant friction between the people who love their country and are trying to do the right thing, and very smart, constantly-conniving sociopaths who feel nothing except for the adrenaline boost they get when they manipulate their coworkers, their underlings, or the nation's citizens at large. These villains often get stolen by other agencies or departments specifically *because* of their reputation for doing whatever it takes to get things done, to include cleaning house of righteous people and violating the law, sometimes flagrantly. Again, sometimes they're caught or the political winds change, and they wind up in prison or at least out the door. More often than not, they wind up getting medals pinned on them. I have worked in several federal and state offices over the years, either as a direct government employee or as a contractor, and I have watched these scoundrels. They are generally brilliant, and the ways they will mastermind ever-evolving plans in the future to spy on and control others will be chilling.

PART THREE:

THE PLAN TO CONQUER US

CHAPTER 7:
HOW WILL THE BRAINPHONE BRING US DOWN?

Do you not know that you are the temple of God and that the Spirit of God dwells in you? If anyone defiles the temple of God, God will destroy him. For the temple of God is holy, which temple you are.
(1 Corinthians 3: 16-17 NKJV)

Brainphone Schemers Will Make Life Without One Very Difficult

Any corporation's dream is for their product to become so woven into the fabric of daily living that people forget what life was like without it. Or, better yet, people can't imagine living a life without it. Televisions, laptops, and smartphones come to mind. Light bulbs are in the mix. Cars, too. If you think deeply about it, however, you might discover that none of these things is necessary for happiness. I'm not sure even one of them is absolutely necessary for living. None of them, I would argue, compares to, say, penicillin or general anesthesia. And yet we live day to day with our electronic devices, nighttime lighting, and expensive, new cars as if they were human organs—a part of us and vital for our existence. Now here's something to consider: Did all of these things instantly become indispensable, or were we somehow lulled into believing that we needed them? Or, perhaps, was life eventually set up for us in a way that we felt we needed them? I don't mean a clever marketing campaign. I mean something more dark and more sinister.

Here's an example. At the paper mill where I worked as a young man many years ago, we had a chemical blending specialist who was old and very set in his ways. He also was frugal to a fault and good at saving and investing his money. He owned a race car, for goodness' sake, and he watched his son race it each week on the local speedway for super-modifieds! He didn't live paycheck to paycheck and, as a result, he often allowed his printed paychecks to pile up for months before cashing them. It drove the mill accountants crazy. They begged him to cash those checks so they could settle up their books.

His reply: "If you don't like it, I'll take cash!"

As it turned out, he wasn't the only person not cashing his paper paycheck at the local bank. There were workers like him throughout the country at different companies doing the same thing and confounding accountants. The solution? Companies began requiring checking accounts. If someone wanted to work for a particular company, he or she was told to set up a checking account in order to receive weekly or bi-weekly pay via direct deposit. Even the U.S. federal government added the requirement. Want a job as a maintenance worker at one of the national parks nowadays? Here's a standard blurb from the job posting: "Selectee will be required to participate in the Direct Deposit Electronics Funds Transfer Program." It sounds tame enough. But think about it. For the bank to handle that direct deposit, it needs the person's Social Security number. If the person's company deposits more than $10,000—say, a yearly bonus—the bank is required by law to report that deposit to the federal government. If that person withdraws the bonus all at once, the bank also is required to report it. Everything, from the transactions to the reporting to the access to the money itself, is electronic. That means hackers can intervene and read about the person's finances or possibility even steal the money. Although it might hold the air of an otherwise innocent requirement, one entity involving someone's paycheck now has morphed into three (possibly four) entities. None of these requirements is necessary: we've just been forced to believe they are.

Put bluntly, because of a bunch of accountants getting steamed about people not cashing their paychecks, we've now been forced into a situation—involving a checking account we've been led to believe that we *really*, *really* need—where our privacy is jeopardized and our money is constantly at risk of being stolen. I'm convinced that these same corporate folks will be the ones, down the road, who tell us that, in order to work for their company, we *really*, *really* need to have a brainphone.

Another example of something we don't need, but that we've been convinced we do, is an online presence. I volunteered a few years ago to teach children's Bible study at my church. The church said that the schedule, announcing who would be teaching which group, would be put out on Facebook each week. I informed the person running children's ministry that I didn't have a Facebook account. "Why not?" she asked, clearly surprised.

"Because," I answered, "I believe Facebook is the playground of the devil here on Earth." And I wasn't joking. Eventually, they worked something out that I would be sent the schedule through my email account. "Crazy old man," they likely were thinking. (Admittedly, there's probably an element of truth to that observation!)

I get lots of surprised looks when I tell people that I'm not on Facebook, Instagram, or Snapchat. If you've read this far into this book, you surely understand why I don't have these accounts. But the surprised looks I get aren't because people can't understand why I don't like something so many other people do: they give me those surprised looks because they can't understand why I don't have something that's clearly *vital* to a person's existence. Social media became essential because companies were able to: 1) convince people they *had* to have it; 2) and drip the dopamine

We've been sold on the notion that we need an online presence. Photograph by John Hoang. Courtesy of the photographer via Unsplash.

in a way that people became hooked. When you think about it, Items 1 and 2 are kind of the same.

But if forcing people to have direct deposit or peer-pressuring them into opening a Facebook account is a bit mean-spirited, then convincing them that they *must* get the latest and greatest brain implant is about as deceitful as it gets. And setting things up in society where people feel they have no other choice is downright fiendish. No matter how quick and painless the procedure is, we're still talking about adding machinery to the fragile, almost jelly-like, matter between our ears that makes us who we are. But the schemers will find a way to convince us that such a procedure—done quickly, cheaply, conveniently, efficiently, and involving one of the most fragile parts of the human body—is something we can't exist without. Once the device becomes essential, so will the applications, the accessories, the upgrades.

Dr. Nicole Vincent, a neurolaw researcher at the University of Technology Sydney, similarly worries about the self-learning aspects of brain-machine interface (BMI) technology. That is, if the equipment is designed to learn about each brain and make adjustments accordingly, how do human technicians keep track of what is happening to their customers? Brain tech consumers, she suggests, may then become alarmingly reliant on their BMI device and whichever tech giant is providing it. "If the device alters itself as it learns about how to communicate with your brain," she warns, "you'd better hope that the device can be reproduced if it breaks, because if it can't, then you would effectively be suffering from something like brain damage, and I really worry about this sort of scenario" (Funnel, 2021).

Imagine you go into a brainphone store at the local outlets. You get a good deal on a new brainphone, and you lie down at the robot station where the device is implanted while you wait. Halfway through the procedure, the electricity goes out at the store, and your robot is frozen in place. You're taken to the local hospital to have the neural threads removed and to have the hole that's been augured into your skull refilled. The fill is temporary, similar to a temporary filling for a tooth. Everyone, including you, assumes that you're going to try again at a later date. It's just a matter of healing for a bit.

About a month later, you give the procedure another go around. This time the electricity stays on, and the robot performs flawlessly. Your brainphone is working. However, a week later, you receive a recall notice in the mail. There is a malfunction in your brand and model of brainphone that's causing people to inexplicably jump off tall buildings. The company is offering you a replacement for free. God bless their little hearts. So, you go back to the outlet, present your recall notice, and have them install a replacement model.

Now you're in business. You tap into The Hive that consists of your family and friends, and you check out some other hives, consisting of people you hope might want to include you. It would be a step up to belong to one of their hives. Uh oh. A problem. A hive you would *like* to belong to won't even consider you, because the newest model brainphone just came out, and the one you have is an older model. They look down their noses at you. Ha ha! It was silly of you to think you even had a chance to belong to their group.

A few months later, you scrounge enough money for the upgrade. (The company gives you a great deal on a trade-in. It makes you wonder: Is your used brainphone going to be inserted in someone else's brain as a used model? Blechhh!) You now have the latest and greatest brainphone there is—the X-27! You approach a few of the exclusive hives. One or two give you the impression that they might—*might!*—let you in someday.

A month or so later, you are now a member of a new hive. Your new friends are snooty and shallow. They're petty. The things they communicate about are meaningless. As with many things in life, you realize that there's a difference between what people want and what they *think* they want. You thought you wanted to be part of an exclusive hive. Now, it hardly seems worth all the time, money, and trauma.

Speaking of trauma, guess what? All those implant procedures, done by the mall-store robots, have caused a bit a traumatic brain injury. You are starting to get mentally confused. You're becoming irritable, aggressive, and impulsive. You have begun slurring your speech (which isn't as much of a problem as it could be, since people with brainphones don't have to speak to communicate). You have head-splitting headaches.

You get checked out by a brain surgeon. She says your brain is bleeding and you need to have an operation. "Can you fix the memory and emotional problems, doc?" you ask.

"Of course," she responds. "There are all kinds of brainphone apps for that!"

You leave the office, realizing that the possibility of removing the brainphone was never brought up by either you or the surgeon, not even in passing. That would have been ridiculous, of course. After all, you *need* your brainphone, right?

Brainphone Schemers Will Take Us to the Event Horizon

An *event horizon* is the theoretical entrance to a black hole, where everything—even light itself—gets sucked in and can no longer escape. It is also at this entrance that whatever is getting sucked in can no longer reach out and touch the person watching the phenomenon happen. In other instances, something might be called an event horizon when a point of no return has been reached. Oftentimes, the observers don't even realize they've reached that point.

We're not quite at the brainphone event horizon yet. This book argues that brainphones will be mass produced in a few years. They will be mass marketed. Artificial Intelligence eventually will take over the network. These happenings will dramatically alter us, eventually destroying us. But, importantly, by the time humans figure out that life without the brainphone is a preferred existence, it will be too late. We already will be enslaved. We will have entered the event horizon, and there will be no turning back.

There is, however, one big difference between a black hole's event horizon and the brainphone's event horizon: In the case of the brainphone, the event horizon is not happening naturally. That is, it is being planned by others. There are people who *want* you to be so far into the brainphone experience that you no longer can break free. At that point, they'll have your money, your mind, and all the data that your mind produces. (The data either will be for sale or used for oppression.)

Are there really people this arrogant? Yes, there are. And among them, in my estimation, are the high-tech billionaires. They are the ones financing the brainphone and its network, and they have grand designs. Science documentarian David Malone says the hubris among this group is frightening.

> If you don't [have a conversation about advanced technology], the conversation is going to be had by a lot of techy boys, who are billionaires, who are going to decide [for themselves] what the future is—because they know better than you. And, having interviewed them, they really think this! (The Institute of Arts and Ideas, 2020)

So, a limited group of very influential people are calling all the shots right now, and they are bringing us to the event horizon without hardly any debate at all on the topic.

So that's the rich. But what about the powerful? At least in the United States, it is possible to be a congressman and not be worth a lot of money. In fact, many very powerful administrators in government departments—even Cabinet administrators in the White House—make less than many plumbers do. Surely, they don't want to see people, as a species, brought to the brink, right? Hmmm, maybe they don't care. Remember, politics is an industry. The product is Re-Election. Your purchasing power for this product is your vote. And the marketing for this product is bent on making you believe, at least in November every so often, that you have the Desire to Re-Elect. Politicians understand that such marketing costs a lot, and so they spend inordinate amounts of time calling very rich people and asking them to contribute. Might there be a high-tech billionaire or two among those contributors? Undoubtedly. And might they be seeking something in return for their contributions? Sure—that's what makes the world go round. (Well, it's what makes Washington, DC, spin on its axis, anyway.)

More importantly, politicians likely will be slow to push back against the brainphone because *they* will see benefits for themselves in monitoring people's thoughts. Namely, their constituents' thoughts. Imagine if you're a career politician and, for a fee, you are able to access the hopes, dreams, and fears of the voters in your district. Aren't you going to be interested? Even people with something close to an ethical compass likely would be entranced with having such information. So, you *know*

Politicians will love the brain data they gather on their constituents and on their enemies.

that most politicians won't be able to say no. Let's take it even further. Suppose that, as a longtime and perhaps somewhat crooked politician, you are able—through unscrupulous channels—to gain the brain data on your political opponent. Or your political enemies (local activists, investigative news reporters, an attentive local prosecutor, etc.). Aren't you going to be rivetted to the idea? I'm guessing yes. Unless we do it soon, it is going to be very, very difficult to get laws against brainphones passed in the future. Politicians will be too eager to tap into all the useful brain data out there to ever give up such a thing.

My guess is that rich and powerful people also will bring us to the event horizon by using something salespeople call the salami slice method. You could never eat an entire casing, or deli-sized stick, of salami. But you could eat it a couple of slices at a time. In fact, that's how most people eat it. The same goes with sales. If you are a salesman trying to garner all of the information technology (IT) business at the local, large corporation, you'll probably have a difficult time muscling out the tech company that currently holds the contract. But, say, like eating a slice of salami, you convince that corporation to give you a small portion of their IT work—maybe the laptop upkeep. Then, when that goes well, you ask for another slice, like, say, the cloud data storage. Slice by slice, you work your way in and push the old tech provider out the door. At some point, you will have eaten the entire stick of salami. The big IT contract with that corporation is yours.

Such it will be with getting us to the event horizon. First, we'll be asked not to have a full implant, but, say, only the neural threads going into our heads. Then, we'll be asked to have the full implant, but only to assist us in memorizing things. Then, for the gaming. Then, for the communication. Then, for the purchasing power via our minds. Slice by slice, we'll become more immersed in life with the brainphone. It will then be too late to pull ourselves out of the black hole.

"Scott," you might respond, "you're not giving enough credit to the human race. Surely if things get as bad as you say they will, people will be mindful enough of their situation to have the devices removed." I have two responses to counter that response: 1) the smartphone; 2) social media. Many young tweens and teenagers with smartphones are destroying their capacities to process information and be happy in this world due to their smartphone usage. Who is yanking away their phones in the interest of saving them? Hardly anyone. And allowing people under the age of 18 to participate regularly in social media is having a similar impact on young people's mental health, particularly in the areas of belongingness, fear of missing out, feelings of inadequacy, and bullying. No one cares. A generation of young people are riding their smartphones and social media accounts right off a cliff, and neither they nor their parents will see the event horizon until they have fallen halfway to the ground below.

Brainphones Will Ruin Our Personalities

Brainphones will dramatically change the social and physical make-up of people who use them, and the implants will alter entire generations and communities. The brainphones will be one with their owners, and the owners will be one with the collective. Another possibility: Unlike the *singularity*, where AI takes over the world, the brainphone network might create a more hideous reality, where AI and humans become forever intertwined.

Remember the movie *Spider-Man 2* (2004) with Tobey Maguire? In the film, Dr. Otto Octavius (played by Alfred Molina) becomes accidentally merged with his mechanical arms, resulting in his transforming into the villainous Dr. Octopus. The long, machine arms, now fused to his spine, have Artificial Intelligence built into them. An inhibitor chip meant to buffer Otto's brain is smashed, and the arms begin to affect Otto's personality, making him even more sinister than he otherwise might have been. That scenario is essentially what some experts are predicting will happen with the brainphone: People's personalities will change from brain-machine interface devices, and they will neither be aware of the change nor will they recognize where the change is coming from. "For some of these patients, these devices become such an integrated part of themselves that they refuse to have them removed at the end of the clinical trial," observers bioengineer Dr. Rylie Green of Imperial College London. "It has become increasingly evident that neurotechnologies have the potential to profoundly shape our own human experience and sense of self" (Frum, 2021).

Another impact the brainphone will have on our personalities is that the diversity of personalities will melt away, essentially forming one, big hive personality. I've referred throughout this book to people—connected via technology—making rash, collective decisions. But what about a collective personality? Seems possible to me. As people become acclimated to their brainphones and begin communicating without talking, the speed of information transference will ramp up considerably. Think about a drummer on a snare drum. She hits the snare with her left drumstick, then her right drumstick. Left. Right. Tap. Tap. Tap. Tap. Each tap is a distinct sound, each processed separately, right? But if she goes into a drum roll, accelerating so that there are several taps in each second, the taps are no longer distinguishable. To the listener, it is one, long hum. Such it will be with information, once people become accustomed to quickening the exchange. Communication will be a collective whirr. Comprehension will be a collective whirr. Personality will be a collective whirr.

Literature, music, and art likely will disappear. What's the point of taking six months to write a book if people in The Hive ingest it and comprehend it in seconds? What's the point of writing a song if The Hive collectively decides whether or not it has a good beat and is worthy of playing, not aloud but silently within The Hive mind's internal jukebox? What's the point of creating a painting if it won't touch each admirer differently?

The hum of, say, hundreds of millions of people in one country will be too difficult to handle. Therefore, people will have to focus on a particular drum set. That's where hives come in. A person's hive will be the brainphone network made up of his or her family and friends. A hive of 50 to 100 people will likely be the norm. Hives might greatly overlap, but it's doubtful, as following many different information hums will be difficult (but not impossible). More likely will be a brainphone humming on two hive channels—Hive Channel One for family and friends, and Hive Channel Two for work colleagues. The collective personality that a person absorbs will change daily as he or she switches from Channel One to Channel Two and then back again to Channel One. Does it sound like people will become zombies? I agree—it does.

Somewhat counterintuitively, I also believe people—hoping for something visceral that their caveman brain longs for—will find themselves living and existing, perhaps zombified in a room, with the people in their hives. Sort of like texting with a person who is sitting next to you on the couch. No talking, no music, no eye contact, and perhaps even no touching. Just a group of people in a circle, immersed fully in the hum of their brainphones as information gets shared constantly among one another at the speed of electricity.

Brainphones Will Ruin Our Communities

Over the generations, people have survived because they have joined and have stayed a part of tangible communities—groups of people, hometowns, churches, companies, and traditions. One huge problem with the Web is that it has unmoored us from these tangible things. Digital sociologist Julie M. Albright calls this condition *untethering*. The word seems a bit ironic: people seem forever tethered to their smartphones,

A hive will be both physical and mental. Just a group of people in a room, communicating telepathically.

and so, the idea that smartphones are, in fact, UNtethering us is something that is tough to wrap one's head around. But here is a good way to understand it: If the Web pulls us away from things that are real—that we can touch—then we are able to dangerously form our own sense of what reality is. Even more destructive: Others can decide for us what we should accept as reality. Writes Albright (2019):

> Living online means that a broader spectrum of our behaviors are traceable and trackable, leaving us open to persuasion, manipulation, surveillance, and control. Our past is never past anymore: digital natives may strive to "live in the now," but the network never forgets. Other negative impacts are surfacing, including declining mental and physical health, waning social skills, and a lack of accountability, which encourages behaviors like "ghosting" on romantic or work relationships, as well as other nefarious exploits.

This condition explains why people these days cannot discuss anything rationally, and why people—particularly young people on college campuses—have become so hypersensitive and anxious. It explains why the political parties are not simply apart, but downright removed—untethered!—from each other's reality.

I once taught a course where the alleged deeds of a famous rap star were the topic of conversation. The consensus in the classroom was that he should do many years in prison for the way he had allegedly mistreated some of his fans. I was asked to offer my approval of this consensus. "I can't do that," was my response. "The man hasn't even been charged with a crime yet, much less tried."

Jaws dropped. Several students looked at me incredulously. Others were angry. They asked me to explain myself. "Well, in this country," I explained calmly, "someone is presumed innocent until they're proven and found guilty. I'm not hearing anything other than hearsay right now based on a few tabloid stories and Web chatter. So, I'm going to play devil's advocate and say that many of you are rushing to judgment."

They were rushing, all right. A few of them rushed out the

Albright warns us of a great societal untethering due to technology. Photography by Greg Willson. Courtesy of the photographer via Unsplash.

door and made a beeline to the school's dean, complaining that I was sanctioning this rap star's alleged actions. Huh? I was stunned by the reaction, but I shouldn't have been. Life with smartphones has created hives where friends within their beeswax home think alike and feel safe and protected by the likemindedness of their brothers and sisters. An outside opinion is the equivalent of an intruder. Adrenaline producing. Physically threatening to them. Frightening to them.

The condition of being untethered also implies forever feeling ungrounded. A good friend of mine, Gary McCabe, is a church pastor who has worked with youth for years and has traveled the world doing missionary work. He notes that young people are becoming increasingly unmoored. "Among youth—even in parts of the world just becoming familiar with connectivity—there's no stake in the ground anymore," he observes. "Nothing sticks for very long. There are no trends, no traditions. There's no time to learn anything deeply before it changes again. Much of social and cultural knowledge has become wide but very shallow."

As a practical matter, it is also difficult to engage in your real-life community when you're always looking down at your smartphone. People who are looking down at their phones aren't really in the place and in the moment, are they? Now, imagine talking with someone who has a brainphone implanted in her head. You're speaking to her, and she's nodding. But is she really listening? Her eyes seem a bit distant. Is she surfing the Web in her mind while you're talking? The nerve!

The town I live in has an annual bathtub race. Yes, you read correctly! The local businesses take old bathtubs and fasten axles and wheels to them. Some of these contraptions are marvels to behold. A person sits inside. Another person pushes the tub from behind. They race, two teams of two at a time, right down Main Street. Through a series of eliminations, the fastest tub team wins.

Sound silly? Maybe. But thousands show up for the event. There's music and fine, outside dining. There's a bathtub parade just before the races. Then the teams are introduced. There's the smell of hay everywhere, as the bales buffer the raceway from the crowd. It is the quintessential community affair. And smartphones don't have a thing to do with it.

Or do they? Sadly, during the races these days, the crowd is peppered with people who are taking videos of the event. They're not watching the race, or cheering, or waving to the person across the way that they recognize. No, they're staring at the event through their smartphone screens, as if the recording—to be uploaded onto social media later in the day, no doubt—is more important than enjoying the moment as it's happening. What will happen when *every* event is experienced through the lens of our brainphone rather than through the here-and-now aspects of the event itself? What about the smells, the tastes, the wind blowing? The Mr. Bubble? Are any of those things registering? Is the brainphone enthusiast really enjoying the bathtub race, or is he concentrating on the taping controls displayed in his mind, checking the framing of the shot, and ensuring the recording volume is perfectly balanced? A minute after a race is over, can he answer anything at all about what just happened? In all likelihood, he would have to press REWIND and replay it in his head to see what he missed!

And so it will go, as community living disintegrates into individuals recording everything and watching them by

Some community events, like bathtub races, need to be enjoyed in person and in the moment. Public domain photo. Courtesy of Wikimedia Commons.

themselves. Life will no longer be lived in the moment or enjoyed in the physical company of others. It will be a flurry of detached snippets, some in the present, some in the recent past, some in the distance past, with little rhyme or reason to them.

Perhaps even worse, if life is experienced in such a singular, disconnected way, there will be little inclination for people to help one another. I have seen firsthand some amazing, neighborly help in my life—people in floodwaters being rescued, neighbors putting out a fire by themselves, people handing out food baskets to the needy in their communities. Yes, people will be "wired" to everyone else. Yet, I can't help but wonder how shared support will change once everyone is living his or her physically-isolated, brainphone existence. Will charitable activity wither away to nothingness?

Brainphones Will Ruin Our Bodies

When I offered my small, public voice to sound the alarm about the upcoming brainphone and its potential harm to humanity, I began with a simple, logical argument called a syllogism, which uses two, true statements and deductive reasoning to reach a solid conclusion. My syllogism:

Major premise: *Smartphone effects are harmful.*
Minor premise: *Implants will magnify the smartphone effects.*
Conclusion: *Implants will magnify the harm.*

Of course, if one were to switch out the words *harmful* and *harm* with *good* and *goodness*, one also would reach a logical conclusion. Which is more appropriate—*harmful* or *good*, or some mix of both? As discussed extensively in Chapter Two, several researchers have supported my major premise (*harmful*). My hope while writing this text was that very smart, influential people might agree with the minor premise—and, perhaps, the conclusion—as well.

As it turns out, while I was working on the last few chapters of this book, a group of 17, well-respected researchers—in fields as diverse as evolutionary biology and technology ethics—published a paper in the prestigious, peer-reviewed journal *Proceedings of the National Academy of Sciences of the United*

States of America (PNAS), essentially offering the same outlook: A next generation of smartphone technology and social media has the potential to destroy humanity. (They also suggest it could destroy the earth as well. Yikes!) Seems a little over the top, right? But the report is very methodological. The paper calls for the study of this technology/media to be treated as a *crisis discipline*, similar to vaccine medicine and climate science (Bak-Coleman et al., 2021). In an interview about the research paper with *Vox*, one of its authors, Joe Bak-Coleman, explains that the direness of the argument stems from the complex and very fragile system that exists among people on Earth:

> The question we were trying to answer was, "What can we infer about the course of society at scale, given what we know about complex systems?"...[Our goal was] to take that perspective and then look at human society with that. And one of the things about complex systems is they have a finite limit to perturbation. If you disturb them too much, they change. And they often tend to fail catastrophically, unexpectedly, without warning. We see this in financial markets—all of a sudden, they crash out of nowhere. (Gaffary, 2021)

In other words, humans exist, interact, and survive as a fragile, complex system. If you knock over a few of the wrong dominoes, a chain reaction starts, and all the dominoes topple. Another of the paper's authors, Carl Bergstrom, says that, rather than simply examine how future communication technology and social media might affect an individual, smart minds need to project the impact of these things onto humanity as an organism. "So what we're saying," he suggests, "is we really want people to look at the large-scale structural changes that these technologies are driving in society" (Gaffary, 2021). Makes sense to me. And when these smart minds determine the magnitude of the harm that's about to happen, they need to go to the nations' leaders and make some vital suggestions.

That's quite a spectrum of possibilities, when you think about it. The harm to your body goes from mild brain bleeding all the way to the end of humankind. And I don't mean "humankind as we know it," although that is a possibility to consider. I mean, the end of humankind as in, "Gee, whatever happened to those creatures called humans that used to roam the earth?"

Another likely, physical side effect is the type that has absolutely nothing to do with the cure or the problem the cure was hoping to remedy. I'm thinking specifically of common Parkinson's disease drugs, particularly those used to treat restless leg syndrome (which is an irresistible urge to move one's legs). The treatment, involving a class of drugs called *dopamine agonists*, was started about 20 years ago. The side effects have been, well, something else. They include a Parkinson's patient's tendency to compulsively gamble (to the point of losing his or her house), compulsively shop, or engage in compulsive sexual behavior, such as masturbation or sex with a partner to the point of injury. This isn't a one-in-a-thousand type of side effect. At least one study suggests that as many as one in seven of the millions of patients taking agonists are experiencing psychological side effects (Moore, Glenmullen, & Mattinson, 2014). Drug researcher Thomas Moore, who conducted the study, notes how alarming the scope and seriousness of the side effects are. "That is a striking psychological side effect rate," he suggests in an interview. "There are a lot of forms of impulse control, but this is a striking and unusual list" (Scott, 2014). And it brings up all kinds of questions regarding the weighing of risks against benefits. It's not much a leap from considering this example to surmising that the brainphone, meant to augment humans in learning, remembering, and communicating better, might create serious side effects—completely unrelated to how the brainphone works or to the brain functions it is meant to improve.

Aside from all the physical side effects from having a foreign object in one's brain—including the oddly related ones—there is a basic health question tied to the brainphone. If you are in a virtual world (gaming, cyber-vacationing, cyber-meeting), and you enjoy a meal in that simulation, is your brain tricked into thinking you've actually eaten? And, if so, does your body respond by thinking it's full? If you drink a bottle of water in your simulation, does your body feel quenched? And the same goes for exercise or sleep—or even breathing. If you feel fit as a fiddle in your alternate universe, but you're withering away in reality, your body soon will be destroyed. Maybe when you die in real life, there will be enough juice left in your brainphone that you may enjoy your cyber-world for a little longer. But since your real self won't be able to pay your bill, even *that* reality won't last long.

A review of all these bad effects summons a worthy question: If it's apparent that the brainphone will bring enormous damage to our autonomy, our personalities, our sense of community, and our physical health, then why are the people planning the device continuing to do so? I have a few theories:

1. As mentioned earlier in this book, wealth and power are addictive. A person might watch his entire world crumble around him, but if he's richer in the process, and if he's in charge of the poop show that's happening, his addiction tells him that's okay.
2. People love experiments, regardless of the outcome. Curiosity forces them to flip the switch, even if the monster

Through brainphone simulation, can your body be tricked into thinking it is full from eating? You might starve to death! Photograph by Pablo Merchan Montes. Courtesy of the photographer via Unsplash.

they're creating with that electricity is likely to turn on them and kill them. They may even *know* the monster is going to kill them in the end, but, like toddlers touching a hot stove top after being told not to, they just can't help themselves.

3. The people controlling the brainphone believe they can remain removed enough from the network so as to not be sucked into it. This attitude is similar to the drug dealer who believes he won't become hooked on his own product.
4. Also mentioned earlier (and courtesy of writers James Cameron and William Wisher, as well as actor Arnold Schwarzenegger): It's in our nature to destroy ourselves.
5. Evil exists in this world.

One theory I don't subscribe to: The people who are inventing this device are torn about its appropriateness, and all they need is a little convincing to stop the designing and the planning. (Maybe by reading this book?) I'm not saying there aren't people who started out at one of these corporations, became horrified, and left. But I'm guessing that they did so a long, long time ago. The decline of humankind is continuing on without them.

CHAPTER 8:
HOW MIGHT THE BRAINPHONE SIGNIFY THE END OF TIMES?

Then another angel, a third, followed them, crying with a loud voice, "Those who worship the beast and its image, and receive a mark on their foreheads or on their hands, they will also drink the wine of God's wrath, poured unmixed into the cup of his anger, and they will be tormented with fire and sulfur in the presence of the holy angels and in the presence of the Lamb."
(Revelation 14: 9-10 NRSV)

Why Consider Biblical Prophecy in a New-Science Book? Do you find religion to be outdated and oppressive? Do you find the teachings in *The Holy Bible* to be quaint but no longer relevant? Do you think old writings are incapable of foretelling the future? Do you think prophecy or a preordained Fate of Humankind is the stuff of silly legend? No problem. Skip directly to Chapter Nine. You will have missed not a beat. No continuity issues whatsoever.

However, if you are of the mind that *maybe* there's something out there more than just dumb luck and the chance arrangement of molecules in our universe, then read on. Jordan Peterson, renowned clinical psychologist and Canadian philosopher—who is painstakingly scientific in all of his teachings—still incorporates Biblical stories in his lectures and writings. Throughout years of interviews, his eventual leaning towards a belief in God and in Christianity is noticeable. Peterson says that one may consider Jesus objectively as a historical figure, or one may consider Him from the perspective of Biblical narrative, which provides the reader with the wonderful, mystical story of Jesus as the Savior of Humankind. Peterson suggests that enlightenment happens—as it did for him—when the two vantage points collide. "Because," he reveals, "I've seen, sometimes, the objective world and the narrative world touch.

Clinical psychologist and Canadian philosopher Jordan Peterson. Photograph by Gage Skidmore. Courtesy of the photographer via Creative Commons. License https://creativecommons.org/licenses/by-sa/3.0/deed.en

You know, that's union synchronicity. And I've seen it many times in my own life and so, in some sense, I believe it's undeniable" (Goins-Phillips, 2021). In other words, it is difficult for even the most skeptical person to step back from oneself, consider his or her life as if it were a giant painting, and not say, "The angels, for whatever reason, have brought me to where I am today."

New-science phenomena and religion are forever tied together. This book is published by Adventures Unlimited Press, a publishing house with a catalog that prominently features both a section on "UFOs & Extraterrestrials" and a section on "Philosophy & Religion." Now, one might think that believing in UFOs and beings from other solar systems might be antithetical to religious beliefs. After all, most religions pertain to Earth, the origins of man, and God's relationship with humans (created "in His own image"). Becoming convinced there are other races in the universe might cancel out that special relationship, right? Furthermore, UFO conferences throughout the world have become increasingly popular over the last ten years, a growth that corresponds almost perfectly to a ten-year, unfortunate decline in religious practice. This book's editor, Jennifer Bolm, and her husband, archeologist and writer David Hatcher Childress, regularly give presentations for packed houses at AlienCon® conferences across the country. At the 2018 AlienCon® convention in New Year City, Kevin Burns, the executive producer of *Ancient Aliens* on The History Channel (a show which, by the way, stars Childress), was asked about the popularity of the show and that of ET studies in general. His response, essentially dismissing the notion of a UFO/God dichotomy, is that people are riveted because (…wait for it…) they are interested in the existence of God! That is, the study of ancient space travelers and the study of religion mesh, allowing people to delve into the mysteries of our existence in a world that has become over-reliant on data. "[*Ancient Aliens* is] not about little green men in outer space. That's the three-headed snake lady that gets you into the tent," Burns tells a reporter from *The*

"God vs. Unexplained Phenomena" seems like a poorly constructed dichotomy. A well-researched person clearly can believe in both.

New York Times. "It's really a show about looking for God. Science would have you believe we are the result of nothing more than a chance assemblage of matter" (Kurutz, 2018).

So, why consider Biblical prophecy as it relates to the brainphone? The answer, in my estimation, is straightforward. If the brainphone, as the research in this book suggests, is leading us very soon to the end of humankind, and if religious dogma foretells the end of humankind, then why shouldn't both be considered in the same text? Perhaps our insight from the doctrine also might offer us some clues as to how we can stop the braintech phenomenon from happening, or at least step aside while others jump off the tech cliff to ruin.

There is no religious creed more compelling when it comes to predicting something like the brainphone and how it fits in with the sunset of man than the Revelation of Jesus Christ—also known as the Book of Revelation, the final chapter in Christianity's *Holy Bible*. Rich with vivid narrative, the chapter arguably was written by John the apostle, exiled by Caesar to the island of Patmos. John is visited by an angel who tells him that he is about to see several prophetic visions, some of them including Jesus. The angel instructs John: "Write on a scroll what you see and send it to the seven churches" (Revelation 1: 11 NKJV).

John sees some visions, all right, and many of them are terrifying. In the future, Satan appoints an Antichrist as the leader of a satanic global government. This Antichrist forces the world to worship him. Then, continuing the chain of delegating, the Antichrist appoints a False Prophet, who deceives the world by performing supposed miracles. The False Profit places "the mark or the name of the beast" (Revelation 13: 17 NKJV) on the forehead or the right hand of those who are willing to worship the Antichrist. Many of those who refuse the Mark are killed as martyrs. The Antichrist and a legion of his worshippers proceed to take over and run a large city of luxury and decadence.

Then, Jesus Christ returns to Earth to avenge the martyrs. Jesus and His army of angels fight the Antichrist, the False Prophet, and their evil army in the Battle of Armageddon. The good guys win. The Antichrist and the False Prophet are tossed into "a lake of fire burning with brimstone" (Revelation 19: 20

NKJV), and their wicked minions are killed by the swords of the angels. The remaining Antichrist followers (that is, the wearers of the Mark) are sent to Hell for their eternal punishment. Saints and true believers celebrate on a new earth without seas and without nighttime, joined together in festivity with Heaven.

The Book of Revelation is rich with extraordinary imagery. Jesus is described as having white hair and fire coming from His eyes. There's the description of the ancient Book of Judgment, with its seven seals, held by the right hand of God. There are the Four Horsemen roaming the earth, each administering one part of God's early sentence on the unsaved: conquest, war, famine, and death. There are the different world events that happen as each seal on the Book of Judgment is broken, such as the blackening of the sun and the reddening of the moon. Then there are the seven angels and the seven trumpets, each rendering the earth with more of God's harsh judgment. For example, the sounding of the fifth trumpet releases a plague of monstrous locusts from Hell:

> [By sounding the fifth trumpet, the fifth angel] opened the bottomless pit, and smoke arose out of the pit like the smoke of a great furnace...Then out of the smoke locusts came upon the earth...And they were not given authority to kill the [unsaved], but to torture them for five months. Their torment was like the torment of a scorpion, when it strikes a man. In those days men will seek death and will not find it; they will desire to die, and death will flee from them. (Revelation 9: 2-6 NKJV)

There is water turning to blood. There are asteroids raining down upon the earth. Yikes! These are disturbing, penetrating descriptions. Some of the imagery is fantastical, some of it is realistic. But perhaps none of it is more compelling, from the vantage point of the technology appraiser, than that of the Mark of the Beast—the description of how those who reject God and worship the Antichrist will physically identify themselves.

Is the Brainphone the Mark of the Beast?

For as long as I have worked on this book, I have been fascinated—as a practicing Christian—by the notion that, perhaps someday, the brainphone will serve a future Antichrist here on Earth as the Mark on the forehead that badges his worshippers. I initially was curious when I noticed that, while the New King James Version (NKJV) of the Bible describes the Mark as being placed "on their foreheads," the old King James Version (KJV) refers to the Mark as existing "in their foreheads." *In their foreheads*. So, brain implant, right? However, upon delving into a book translation of the Codex Alexandrinus at the British Library in London—a fifth-century Greek, Christian manuscript—and the Codex Ephraemi Rescriptus at the National Library of France in Paris—another fifth-century Greek, Christian manuscript—it became apparent that the correct translation is "on their forehead" (Tregelles translation, 1849). So, not necessarily an implant.

Let's consider the entire passage, translated from the two ancient Greek texts just mentioned:

> And he [the False Prophet] causeth all, the small and the great, and the rich and the poor, and the free and the bond, to receive a mark on their right hand, or on their forehead: that no one be able to buy or sell, save he hath the mark, the name of the beast, or the number of his name. Here is wisdom. Let him that hath understanding count the number of the beast: for it is the number of a man; and his number is 666. (Revelation 13: 16-18, Tregelles translation, 1849)

Powerful stuff, and very visceral in its portrayal. A machine-like device or a machine-readable mark, placed as if by futuristic branding iron on the human right hand or forehead, that people must have in the future in order to engage in day-to-day commerce. The physicality and the straightforwardness of the description—to include something that, frankly, sounds like a brand name—seems very modern-day. To the casual observer, it sure sounds a lot like the brainphone.

Also, the evolution of positioning the brainphone within the human skull might give a devout Christian pause. When a brain-machine interface (BMI) device was first proposed some years ago, the notion was that it would be implanted at the back end of the optic nerve, through the back of head, where that long nerve meets the visual cortex. When such a proposed device reached the design stage, it was initially on the side of the head above the ear and not completely embedded. But things have evolved. At least one device currently being considered by the U.S. Food & Drug Administration is a little further up towards the top of the head—fully augured into and flush with the surface of the skull—and perhaps half an inch, if that, towards the front of the head. See where I'm going with this? Following the path as if it were an inverse exponential curve on a chart, one can imagine the brainphone eventually winding up front and center on the forehead.

Part of the implant's proposed charm, as currently designed by at least one corporation, is that it is hidden under the skin and hair. But I'm not so sure American consumers, even in the near future, will want it to stay there. After all, people don't buy sports cars and keep them hidden. The same goes for $500 sneakers. So

The brainphone might become a fashion item, visible in the center of the forehead, with a company logo on it.

why wouldn't a person investing a lot of money in the latest and greatest technology (until it is someday offered for free) want it displayed on his or her forehead, sticking out slightly, with the company's logo showing? Something of a fashion statement; something of a status symbol.

Admittedly, there are important reasons for calmer heads (brainphoneless heads?) to avoid sounding the Biblical alarm against BMI technology. For example, Revelation enthusiasts have been warning about technology as a precursor to the End of Times for decades. I recently enjoyed reading the meticulously researched master's thesis by Reformed Theology Seminary student Ian MacIntyre. It is an academic, professional-grade consideration of the chronology of Revelation. At the conclusion of his thesis, MacIntyre argues: "The appearance of the Antichrist could occur at any time and is probably imminent given the existing technological capabilities to implement the mark of the beast" (MacIntyre, 2013). Sounds pertinent, right? One small problem—it was written nearly ten years ago. Granted, ten years is a small amount of time when it comes to the history of humankind, but how imminent is imminent?

"Creating fear really isn't useful." So says Dr. Brian Moss, head pastor of a church in Maryland and director of the DREAM church conference, which helps other pastors throughout the world build healthy congregations. He holds a Doctorate of Ministry in Leadership from Liberty University and a Master of Divinity degree from Southwestern Baptist Theological Seminary. His research is concentrated in what's called *textual criticism* of the Bible, which looks at the ancient documents written by hand and helps establish which are the original or most authoritative texts. "The thing that most theologians have done when attempting to understand Revelation is go 'Huh!!!'," he observes, suggesting they are more perplexed than anything. "However, the good news regarding the Mark of the Beast is that the Bible is straightforward regarding its intention. That is, people won't mistakenly take the Mark or be tricked into taking it. When people accept the Mark of the Beast, they will be freely acknowledging that they're rejecting God and worshipping the Antichrist. They will be knowingly doing so" (B. Moss, personal communication, September 2, 2021).

Therefore, although the brainphone *could* possibly be the Mark of the Beast *someday*, a couple of really cataclysmic things would have to happen between now and then to make it so:

1. The Antichrist would have to appear on Earth, appointed by Satan.
2. People would have to know he was the Antichrist.
3. A false prophet, appointed by the Antichrist, would have to offer the brainphone and a similar, hand-implant device to anyone who wanted one.
4. People taking the device would have to acknowledge that they were accepting the brainphone or its hand counterpart specifically to revere the Antichrist.

This scenario prompts a few outrageously hypothetical questions. (For the only-marginally disposed, you might be finding this entire chapter outrageous!) If a person purchases a brainphone in the next couple of years, and the use of that device morphs into being the Mark in future years, may a devout Christian remove the implant in the name of renouncing the Antichrist? If the brainphone affects a person's actual free will (as the research in this book suggests it will), then how can someone freely remove it to show his or her loyalty to Jesus Christ? Furthermore, what if, once the Antichrist appears on Earth, our brain neurons are so intertwined with BMI technology that we can't remove it without killing all brain function? I worry about these things.

Church pastor Gary McCabe offers an uncomfortable scenario—that of the brainphone being used to produce phony religious epiphanies. "Suppose whoever is controlling the brainphones at the time of religious turmoil starts implanting cryptic messages or mirages of miracles into people's minds," he speculates. "People might act in very misguided ways, thinking that they're doing so because of Divine Providence."

However, referring back to Dr. Moss's conceptual framework, it is, indeed, comforting to know that God is unlikely to relegate people to Hell who aren't freely rejecting Him. It eases the fear.

Is the Smartphone the Mark of the Beast?

The short answer is no, the smartphone currently in your pocket is not the Mark of the Beast. The events leading up to the Mark—including people selecting it as a way to announce their loyalty to the Antichrist—clearly haven't happened yet.

Speaking metaphorically, however, it is worth asking: Has the smartphone softened us up to whatever Mark we might be approached with in the future? It reminds me of the joke where the doctor tells an elderly woman, "Your husband has a rare skin condition that someday could kill him. But there are some preventative measures. For example, his skin has become super sensitive, even to the small amount of radiation emitted from that old television of yours. Seems to me you're going to have to get rid of it?"

"Get rid of my television?" she asks.

"Yes."

"And miss *The Price is Right*?"

"Yes, ma'am."

She goes back to her husband, waiting in the car. "What did the doctor say?" he asks.

The woman responds: "He said you're going to die!"

I'd like to think that we're not *that* far along in our collective addiction to technology, from television to laptops to smartphones. But, then again, I never thought I'd see the day where 10 percent of people surveyed would admit to checking their smartphones during the act of sex (SureCall Company, 2018)!

Catholic Online wrote a commentary about seven years ago called, "You Already Carry the Mark of the Beast." The writers basically argue that we have become so mesmerized by our smartphones that they might as well be the Mark of the Beast, as we could never give them up, no matter the punishment that might await us at The End. The site says: "To round out the mark of the beast, our phones and mobile devices are becoming payment centers…Doomsayers like to predict that we will all be forced to wear the Mark of the Beast someday in order to do business, but the fact is that we already carry it with us wherever we go, a mark we have willingly accepted for ourselves…It is already possible to buy and sell everything with just a swipe of our phone. Cash and cards are becoming obsolete" ("You Already Carry," 2021). So, we didn't miss the boat, and the smartphone itself is not the Mark of the Beast. But perhaps we're in too deep, technologically speaking, to renounce the Mark once it actually presents itself.

The television series *Star Trek: The Next Generation* has an episode called "The Game" (Season 5, Episode 6), where the crew of the *Enterprise-D* begin playing a virtual game in their minds—courtesy of head-held devices provided to them by the crafty Ktarians. Within your mind, you can see spinning plates and several cornucopias. Every time you manage to spin a plate into a cornucopia, your brain receives a very, very pleasurable shot of dopamine. (I have to give it to the writers of *Star Trek*: over the decades, they have been very much in tune with where this world is headed. This episode aired in 1991, long before online gaming, smartphones, and brainphones.)

The crew becomes addicted to the game, practically incapacitated, and about to be taken over by the Ktarians, who knew all along the device would have such an effect. It takes the young, non-playing and non-addicted Starfleet cadet, Wesley Crusher, to repair the sabotaged android Data and save the day.

The episode has some pretty intense moments. The captain is addicted and pays no mind (pardon the pun) to Wesley's warnings. Wesley's mother, the normally responsible Dr. Beverly Crusher, sabotages Data (who, as an android, is unaffected by the game) to keep him from taking the game devices away. At different points, Wesley and his girlfriend, Robin Lefler, are held down by crewmembers and forced to play the game and become hooked. In Wesley's case, for goodness' sake, Captain Picard and Wesley's mother are on hand! "Hold him steady," orders Picard.

"It's okay, Wesley. It won't hurt. You'll like it!" insists his mother ("The Game", 1991). (Fortunately, Wesley already has repaired Data at that point, and so the ship still will be saved.)

More disturbing are the questions the episode asks. Can a relatively innocent looking device control our lives? Can addiction, even to a natural substance like dopamine, make us do horrible things to the ones we love? Can technology make us turn violent against our family members and neighbors? We couldn't *really* ever get to a point in our history where people are held down and forced by their family members and neighbors to receive the brainphone. Could we?

Within the context of this chapter, it is also worth asking: Can a device, in and of itself, be evil? "No," asserts Dr. Brian Moss. "Devices aren't evil. They can be used for evil purposes. However, I must add, some technical devices have an extraordinary capacity for evil use" (B. Moss, personal communication, September 2, 2021). (Incidentally, Dr. Moss, along with his noteworthy academic accomplishments, is a former computer engineer.) One obvious example: devices that access the Internet, where young children—when adults are not around—can then easily access the most vile pornography and images of violence in existence. It is certainly difficult to discuss the Internet without discussing morality.

If There Is a Brainphone Prophecy, How Does It End?

In recent years, most Christians I've had conversations with are convinced that we are fast approaching the End of Times. It

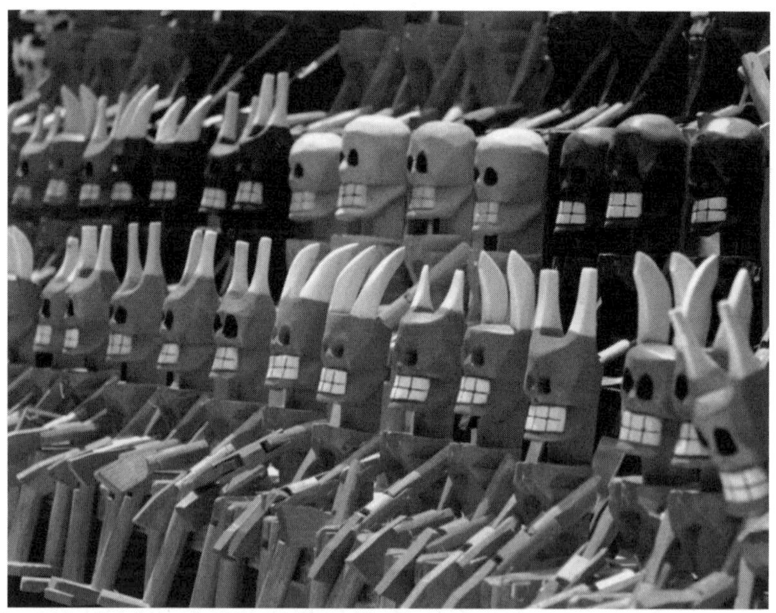

Can a device, in and of itself, be evil? Photograph by Chela Bonky. Courtesy of the photographer via Unsplash.

doesn't have much to do with whoever is president of the United States. It's not the (very real) decline of religious worship in the U.S. And, oddly enough, it is completely unrelated to the COVID-19 pandemic. What has them thinking the end is nigh? I best can describe it as the notion that the world collectively has lost its mind.

Backing away from Revelation, here is an excerpt from the Gospel of Matthew that serves as the source of many of these apocalyptic conversations. The words are those of Jesus Christ, after being asked by his disciples how they will be able to recognize the End of Times preceding His return to Earth:

> For nation will rise against nation, and kingdom against kingdom. And there will be famines, pestilences, and earthquakes in various places...And then many will be offended, will betray one another, and will hate one another...And because lawlessness will abound, the love of many will grow cold. But he who endures to the end shall be saved...Therefore, when you see the abomination of desolation, spoken of by Daniel the prophet, standing in the holy place, then let those who are in Judea flee to the mountains...But of that day and hour, no one knows, not even the angels of heaven, but My Father only. (Matthew 24: 7-36 NKJV)

So, while Jesus says no one knows for sure when the Apocalypse actually is going to happen (including Michael the Archangel and his vast army of angels at Jesus' ready), He suggests that there are signs we can look for so that we might be somewhat prepared. It is among these clues mentioned that friends and acquaintances of mine find reasons for their apprehension. Wars; countries taking over other countries; a megadrought in the United States and a drought, in general, worse than it has ever been throughout much of the world; climate change bringing revolting insects to parts of the world that have never seen them before; a culture of hypersensitivity, constantly finding offense in what others say and "cancelling" them from groups; betrayal in government and in business; extreme hatred between disagreeing groups or political parties; a general disregard for laws; the disappearance of a sense of community; young people losing hope; selfishness; chaos; coldness where love used to be among people. As a good friend of mine once said, "We may not know the day or the hour, but things sure seem a lot ENDier than they used to!"

Perhaps, too, is the sense that, as political correctness becomes so hyper-driven that people are having difficulty expressing their opinions these days, simply holding religious beliefs is becoming politically incorrect as well. I'm not referring to sensitive topics like religious dogma pertaining to LGBTQ+ lifestyles or whether or not having an abortion is a civil liberty. I'm talking about statements as innocent as "I believe in God, and I consider Jesus Christ to be my Savior." When people are afraid to utter something as basic as that statement at work for fear of being fired by a corporation keeping things as bland as possible, it is no wonder why people think there are dark times ahead.

Such sentiment certainly leads to the question: What happens when we all have our brainphones humming, we're all transferring knowledge, feelings, and communication at super speeds, we're all at one with The Hive—but some people in The Hive believe in God and some people don't? Does the majority prevail? Will science rule out faith? Will there be religion-based hives? Skepticism-based hives? Will the government decide for us what is and isn't right for hives to collectively believe? If one hypersensitive, but strong-minded, person within his hive doesn't want religion discussed at all, will the rest of his hive comply so that there is no static? These are important questions for a Christian, who doesn't want to be oppressed, to ask before he or she says yes to the brainphone.

Within the previous excerpt from the Book of Matthew, Jesus refers to the Book of Daniel in the Old Testament, where Daniel the prophet offers a vision of Jesus conquering the Antichrist, similar to that of John in Revelation:

> The fourth beast shall be | A fourth kingdom on earth, | Which shall be different | from all other kingdoms | And shall devour the whole | earth, | Trample it and break it in | pieces… | And another shall rise after | them; | He shall be different from | the first ones, | And shall subdue three | kings. | He shall speak pompous | words against the Most | High, | Shall persecute

> the saints | of the Most High, | And shall intend to change times and law... | But the court shall be | seated, | And they shall take away | his dominion... | Then the kingdom and | dominion, | And the greatness of the | kingdoms under the | whole heaven, | Shall be given to the people, | the saints of the Most | High. | His kingdom is an | everlasting kingdom, | And all dominions shall | serve and obey Him. (Daniel 7: 23-27 NKJV)

And so, in Daniel, Matthew, and Revelation, a general prediction emerges: When the End of Times is upon us, things are going to get a whole lot worse before they get a whole lot better. But for the true believers, it will have been worth the wait and the agonizing persecution.

So, if the brainphone is *possibly* a sign that the end is on the horizon, how does it all end? My thinking is that if we look to scripture as a concern, we also look to scripture for our consolation. If the End of Times is to happen within our lifetime or the lifetime of our children, the story ends with evil, non-believing people (including the non-believing pushers of harmful technology wares) being relegated to Hell. True believers—no longer reliant upon technology and able to think and live freely without implants—watch as Heaven and Earth come together for the party of all parties. My prayer is that anyone who finds himself or herself caught in the Great Tribulation might have the strength to endure some extremely challenging, uncertain times.

Why Bother If It Has All Been Said and Done?

As I've written a few times already, I'm a big fan of the book *The Coddling of the American Mind*, by Greg Lukianoff and Jonathan Haidt. In fact, I think it does an amazing job explaining—in very clear, well-researched terms—why a generation of young people doesn't care about driving, moving out of their parents' houses, or getting married. I bring the book up here because there is also a religious aspect to this phenomenon: If an idle mind is the devil's playground, then the i-Generation is offering Satan quite the large playground, indeed.

So, here's a question: Is the smartphone, combined with social media, conceivably the work of Satan? Or is he just taking advantage of a perfect but time-random storm of technology and social events? I put this question to Dr. Brian Moss. His response: It's the latter. That is, Satan doesn't impose himself into our reality, but he is very creative about using whatever society provides to get people to turn away from God. "Satan uses cultural context to advance his agenda," says Moss. "Taking whatever sorts of advancements are happening at any particular time in our history, he is able to use them to deter people from God's purpose. He is also able to manipulate these advancements in ways that make people feel defeated and disabled in their spiritual journeys" (B. Moss, personal communication, September 2, 2021).

Incidentally, the discussion of Satan as a real entity might be something new for you. Many people today view the devil as a literature-related personification of evil. Or perhaps they look at him as ancient folklore without substance. Or perhaps they look at him as the goofy belief of overly religious people. (Comedian Dana Carvey as the Church Lady: "And who could have led us into that chasm, Jimmy…Could it be—SATAN?") Of course, a worthy discussion regarding the existence of Satan probably could fill another book (or hundreds of them). But it warrants asking briefly here: Is evil personified? Does it plot against us? Does it celebrate when we feel weak, when and hopeless?

For me, the answer is straightforward. As a professional data analyst, I cannot look at the way good things in this world have come together over the millennia and say, "Oh, that's by chance." It doesn't compute. So, with a touch of faith added into my calculations, I conclude there must a God—an intelligent designer who also loves us and keeps an eye on us. Conversely, as a professional data analyst, I cannot look at the evil in this world and say, "Wow, *that* was random!" The ills of this world, in fact, don't strike me as random at all. Most of them seem very contrived. Very strategic. And the enemy is in it for the long haul.

But if Satan exists, why, for now, is he hiding himself? What's in it for him to stay in the shadows? If you consider the lifelong choosing between right and wrong as spiritual warfare, then it makes sense why the enemy stays hidden. In combat, camouflage is everything. So is secrecy. British writer C.S. Lewis—best known for *The Chronicles of Narnia*—was an

Does Satan exist? And, if yes, what role does he play in harmful technology? Photograph by Razvan Chisu. Courtesy of the photographer via Unsplash.

Oxford college professor and a proclaimed atheist who eventually became a passionate Christian. In his novel, *The Screwtape Letters*, elder demon Screwtape explains to his nephew (and mentee), Wormword, why it is best to stay hidden.

> That question, at least for the present phase of the struggle, has been answered for us by [Hell's] High Command. Our policy, for the moment, is to conceal ourselves. Of course, this has not always been so...But in the meantime we must obey orders. I do not think you will have much difficulty in keeping the patient in the dark. The fact that "devils" are predominantly *comic* figures in the modern imagination will help you. If any faint suspicion of your existence begins to arise in his mind, suggest to him a picture of something in red tights, and persuade him that since he cannot believe in that (it is an old textbook method of confusing them) he therefore cannot believe in you. (Lewis, 1942)

In Twelve-Step meetings, alcohol and substance addicts are taught that addictions are so sneaky that they essentially should be personified. The addiction itself is called cunning, baffling, and powerful. I would say the same modifiers apply to Satan. Only, he isn't imagination personified for an addict's hope to recover. He's the real deal.

Admittedly, researching what scripture has to say about Satan and the End of Times brings about a bit of hopelessness and helplessness. Aside from planning to avoid the Mark of the Beast and proclaiming one's personal salvation, what's a Christian to do? That is, let's say, hypothetically, that the brainphone is, perhaps, a foretold event and a part of the sad story of the end of humankind, and, therefore, a component of the End of Times. Why not just throw up our hands and let it happen? Why should someone like me, for example, bother writing articles and a book to warn people? Why should I bother to write to Congress to get a law passed against BMI implants? It's like that line from the song "Destination Unknown," by the group *Missing Persons*: "You ask yourself | When will my time come? | Has it all been said and done? | I know I'll leave when it's my time to go | 'Til then I'll carry on with what I know." Can someone subscribe to religion and fatalism at the same time? Wouldn't it just be better to head for the hills and live a quiet, secluded life while the fight between good and evil plays itself out and comes to a head?

It is interesting that Jesus and the famous philosophers have something in common regarding fatalism. (Religious or not, any reader of the Bible has to acknowledge that Jesus was the ultimate philosopher.) All of them suggest that we must press on and freely do what's right, even though some things are preordained. Jesus says to us, "You are the salt of the earth…You are the light of the world" (Matthew 5: 13-14 NKJV), seemly suggesting that we should press on in our attempts to change things for the better. And, certainly, *The Holy Bible* is the bedrock of free-will philosophy, as we repeatedly are encouraged to publicly and freely dedicate our lives to Jesus Christ in order that He might be our Savior.

Berlin-based philosopher Steven Cave wrote an interesting piece for *The Atlantic* a few years ago entitled, "There's No Such Thing as Free Will, but We're Better Off Believing in It

Anyway." Ha! Great title. Cave suggests that, as scientists look further into how the brain makes us behave certain ways, it seems less and less like we have free will. On the other hand, what scientists *also* are discovering is that the more humans tend to believe in free will, the better off they become individually, and the more society improves. Cave admits how perplexing these two findings sound next to each other:

> Our codes of ethics, for example, assume that we can freely choose between right and wrong. In the Christian tradition, this is known as "moral liberty"...[But the] sciences have grown steadily bolder in their claim that all human behavior can be explained through the clockwork laws of cause and effect...[Still, the] kind of will that leads to success—seeing positive options for oneself, making good decisions and sticking to them—can be cultivated, and those at the bottom of society are most in need of that cultivation. To some people, this may sound like a gratuitous attempt to have one's cake and eat it too. And in a way it is. It is an attempt to retain the best parts of the free-will belief system while ditching the worst. (Cave, 2016)

So, again, the philosophy seems to be, "Yeah, you can't control everything. And, maybe, you can't control *anything*! But you gotta try, if the plan for humanity is going to work."

Perhaps this chapter should end the same way that my extensive interview with Dr. Brian Moss ended—with a thought about seeking change when the sad story seems already to have been written. Dr. Moss instructs us: "We can insulate ourselves from the evils of the world. But we shouldn't *isolate* ourselves from them. We should push for ethical improvements. We should strive to be cultural agents of change" (B. Moss, personal communication, September 2, 2021).

Thus, that's the reason why we should bother to advocate for our kids to be free from man-made circuitry in their heads. Because maybe it hasn't all been said and done just yet. Maybe the fact that we protest is part of the foretelling.

PART FOUR:

OUR REBELLION AGAINST THE TAKEOVER

CHAPTER 9:
HOW MIGHT YOU DETERMINE THAT THE BRAINPHONE IS A BAD IDEA?

So they worshipped the dragon who gave authority to the beast; and they worshipped the beast, saying, "Who is like the beast? Who is able to make war with him?"
(Revelation 13: 4 NKJV)

You Should Recognize that a Storm Is Coming

I would say that the people I've discussed the brainphone with fall into four categories: 1) people (mostly younger) who think the brainphone is about to happen and are, in fact, excited about the technology; 2) people who think the brainphone is real, and it's bad, but it's so far off in the future that there's no point in worrying about it now; 3) people who think the entire notion of the brainphone is a silly work of science fiction; 4) people who think the brainphone is real, it's bad, and it's here—that is, people who have the same concerns that I do. I would say that Group 1 is the largest, decreasing in size down to Group 4. My hope is that writing this book and gently pushing for change might be the small stone that starts ripples in the pond. I would like to see Group 4 become the largest group. I would like to see laws passed throughout the Free World banning the brainphone.

If people recognize the brainphone as a here-and-now threat to our existence, I believe we can push back against total interconnectivity. By recognizing the coming storm, fighting the disease of technology addiction, saying no to corporations and governments, and gaining a new appreciation for personal privacy, I believe we can embrace a lifestyle that keeps brain matter and technology physically separated.

If you are part of Group 1, 2, or 3, how might you change your mind and decide that a brainphone storm is on the horizon? I suppose the first thing you might do is increase the number of sources you get your news from. I'm not sure who is going to support the manufacturing and sale of brainphones, but whichever side does, the other side is likely to oppose it, in the name of hype and good television ratings. Add to your news arsenal a few reporters on the left, a few on the right, and a few in the center. So, if you currently watch Fox News, occasionally check out MSNBC. And vice versa. Add a quality, neutral source, like BBC America. Maybe a religious news outlet. Check out a fringe source every now and then, like *Nexus* magazine. Keep it mixed: don't get too drawn in or mesmerized by any one reporter, commentator, venue, or philosophy.

Next, consider how fast things are changing around you. Technology in particular. Is the change accelerating? Is anyone able to catch up? The circumstances may be more about what's going on around you than you yourself. Such a predicament was forecasted over 50 years ago in Alvin Toffler's book *Future Shock*. He suggested that different types of human events were about to collide in such a way that people wouldn't be able to handle them. He also suggested that the changes would be so temporary that no one would be able to adjust to them, unless, as a species, we somehow made grand preparations for how we were going to absorb (or resist) such changes. Here's one of my favorite quotes, written, coincidentally enough, just after a subchapter on human-robot merging:

> This massive injection of speed and novelty into the fabric of society will force us not merely to cope more rapidly with familiar situations, events and moral dilemmas, but to cope at a progressively faster rate with situations that are, for us, decidedly unfamiliar, "first time" situations, strange, irregular, unpredictable. (Toffler, 1970)

Toffler suggests that some solutions might include embracing life patterns whenever possible; group sharing of best coping practices; better education in adaptation techniques; and pushing back on technological advances when such advances become immoral. (Amen, brother.)

Another step in this self-assessment: Ask yourself what sort of changes you've made recently in your life that essentially give in to technology. How much information about you that otherwise would be private have you placed (or allowed to be placed) on the Web over the last few months? Does the access to this information by people, corporations, and governments make you uncomfortable?

Ask yourself how much time you lose each day due to technology. How much time do you spend socializing on Facebook, Snapchat, Instagram, or Discord? How much time do you spend texting? If you do these things on your smartphone, your phone will provide you statistics on how long you've been on it each day. More than three hours? Hmmm, that's a problem, for all sorts of reasons, including social and psychological.

Ask yourself if you believe corporations have your best interests at heart. Have things gotten any better than when tobacco companies tried to counter information about lung cancer? Or when oil companies tried to counter information about global warming? If the brainphone is, at some point, conclusively shown to be destroying our capacity to function collectivity or as individuals, do you believe the tech companies supplying them are likely to stop doing so?

Along the same lines, ask yourself if you believe your federal government has your best interests at heart. Or have we "fed the beast," as they say, so much and for so long that it can't stop eating (our taxes), growing, and controlling?

Furthermore, ask yourself if your federal government is more involved with your life these days than it used to be. Has it become interested in your thoughts and movements, rather than simply leaving you, as a law-abiding citizen, to just live your life and do your thing? The U.S. Constitution's Fourth Amendment is supposed to keep the government from collecting unlimited information on us (calling such tactics "unreasonable searches"). Instead, a "warrant" needs to "issued" by a judge, and the thing to be sought after should be "particularly described," related to some "probable cause." The notion that this Amendment is being followed these days is laughable. It is looked at as if its ideals were quaint but no longer practical.

Ask yourself if you have seen brain-tech implants mentioned in the news recently. How are the people giving these demonstrations portraying these implants? As something normal? As something harmless? As something everyone should

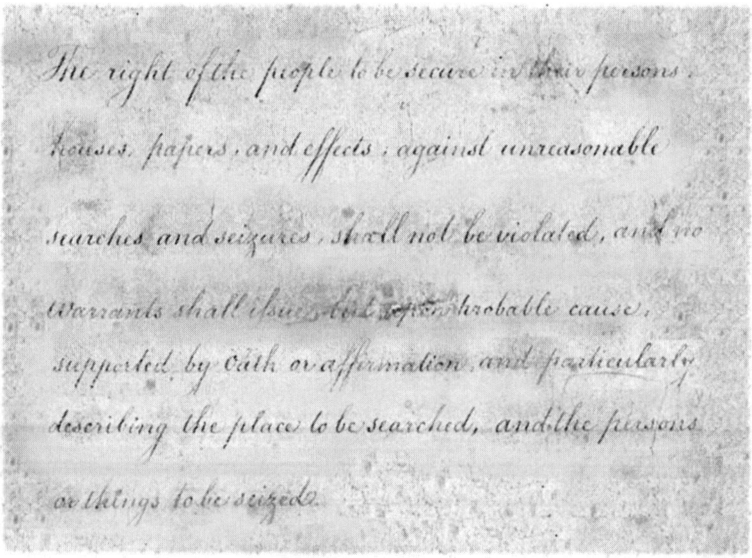
The right of the people to be secure in their persons, houses, papers, and effects, against unreasonable searches and seizures, shall not be violated, and no Warrants shall issue, but upon probable cause, supported by oath or affirmation, and particularly describing the place to be searched, and the persons or things to be seized.

The U.S. Constitution's Fourth Amendment seems largely ignored these days, as if its ideals were considered quaint but no longer practical. Images from the U.S. National Archives and Records Administration.

have? As something that will fulfill your wildest dreams? At this writing, Elon Musk's Neuralink Corporation is the only company providing such exhibitions. But there will be more. One day soon, the advertising and media buildup by companies promoting their brainphones really will be something to watch. It will draw us all in, keep our attention, and get us all talking. Our children will be screaming for brainphones.

Next, ask yourself what addiction has done to your family or the family of a friend. How destructive was it? How much did it suck the life right out of that family unit? Will brainphone addiction do the same thing to you or your family? As mentioned previously, science commentators suggest that the addictive nature of smartphones and social media will be exponentially magnified when metal threads and electrodes become enmeshed in brain tissue. Are corporations worried about brainphone addiction? I doubt it. In fact, I suspect that they're counting on it.

Finally, if you see harmful things happening in your life—right now—because of technology and the Web, ask yourself if you think things are about to become dramatically *more* harmful when this technology is inserted in your brain. As you might imagine, I believe this last question is rhetorical only.

You Should Identify the Enemy

Not everything in this world can be broken down into *good* and *evil.* Although I tend to believe that people are inherently self-serving, I also believe that there's a general heartbeat of goodness in communities that pumps kindness where it's needed most. Many times, bad things happen when otherwise good people are misinformed. Sometimes, a good intention morphs into a bad act. And, certainly, people are complicated creatures. We all are trail-mix cannisters of good and bad deeds.

But, sometimes, intentions are clearly bad. The plan is for something harmful to happen. And the people making those plans are crafty and underhanded. They are very good as setting things in motion and then watching from the shadows while all hell breaks loose. I'm not into "calling people out." After all, today's callout culture is, in part, amped by much of the technology I assess in this book. But if you decide that the brainphone is a terrible thing, potentially bringing harm to you and your family, then it might be a good exercise to pan out, consider the big picture, and consider who might be behind this injurious technology. It might be productive to figure out who the enemies are in all of this. After all, it is impossible to oppose something if you don't know what (or whom) you're opposing.

My first vote goes to corporations, particularly the tech companies, who relish having as much information about you as possible. The more corporations know about you, the more they know about how to get you to buy their very expensive toys—and how to get you to borrow money and mire you in debt in order to buy more and more of them. The desire for money is not, in and of itself, a bad human trait. But, when hidden behind the mask of a corporation, greed goes in some pretty sinister directions. With the brainphone, the greed will be rampant, and the only goal will be sales, sales, sales. As mentioned before, if corporations determine that brainphones are physically and mentally damaging people, they will bury that information. I could use the tobacco industry as a comparison, but that example is obvious. How about something not quite so obvious? How about sugar? In the mid-to-late 1960s, the sugar industry secretly funded research suggesting that saturated fat and cholesterol—rather than sugar—was the main cause of obesity and coronary heart disease (Kearns, Schmidt, & Glantz, 2016). The industry's scheming worked: U.S. government dietary guidelines, for decades, emphasized low-fat foods, even though many of those foods were bursting with sugar and damaging transfats. A British scientist, John Yudkin, tried to refute the research and sound the alarm on sugar, but food corporations worked successfully to ruin his reputation (Leslie, 2016). (It has taken nearly 50 years for his research and scientific life to gain a newfound respect.) Over that time, people grew dramatically fatter. By the end of the 20th century, heart disease was the most common cause of death in the United States. I don't hold people unaccountable: cutting back on sugary foods and exercising should be ways of living for everyone, and many people simply don't do what they know they should. However, it is worth pondering how many lives could have been saved or improved were it not for an industry's clever media and research-funding campaign. Corporations that offer things for our consumption tend to emphasize turning a profit above all else, humanity be damned.

And such it will be with the tech companies, who continue on with their business plans in spite of their own research showing how much harm they're doing. In September of 2021, The *Wall Street Journal* published a damning investigation piece suggesting that Facebook, through its (Facebook's) own internal, three-year study, knew that its Instagram app was a vehicle for body shaming, and that it was psychologically harmful to teenagers, especially teenage girls (Wells, Horwitz, & Seetharaman, 2021). On one internal slide, Facebook's research indicated that 13 percent of British teens and six percent of American teens considering suicide related those toxic thoughts directly back to Instagram.

And yet, unbelievably, at this writing, Facebook is still moving forward with plans to offer a child version (i.e., preteen version) of Instagram (Hatmaker, 2021). (Update: At press time for this book, Meta [Facebook] was pausing the release of the child app due to public outrage and pressure from Congress. However, the plan still exists, as the company—in its press release announcing the pause—insists, "We believe building 'Instagram Kids' is the right thing to do" [Mosseri, 2021].)

My second vote for the enemies list goes to governments, who also desire to hold as much information about you as possible. The more the government knows about you, the more it can control you, and the more it is able to keep its corrupt machinery running without disruption. I'm not an anarchist: I believe a healthy government is good for everyone. But it's a government that needs to be kept in check by its citizens. In his final novel of the *Dune* series, called *Chapterhouse*, Frank Herbert suggests that it isn't simply the condition of absolute power corrupting absolutely, but the tendency of government to attract really bad people:

> All governments suffer a recurring problem: Power attracts pathological personalities. It is not that power corrupts but that it is magnetic to the corruptible. Such people have a tendency to become drunk on violence, a condition to which they are quickly addicted. (Herbert, 1985)

What could be more alluring to such a villain than the ability to read the thoughts of his or her subjects? Is the government going to push you to get a brainphone? Oh yes.

Next on the prospective enemies list: People who are addicted to wealth and numb to their condition. I don't have a problem with people who believe they are entitled to their money. I say God bless 'em. However, I *do* have a problem with people who believe they are entitled to *my* money. These are the types of billionaires and near-billionaires I can see getting everyone hooked on the brainphone solely for their personal gain. I've only ever met one tech billionaire, and that guy had "sociopath" written all over him. If your average sociopath works in a small company, he might bring down the company with his charm, deceitfulness, manipulation, and irresponsible risk-taking. In the end, he'll somehow manage to walk away with lots of money, and everyone else will walk away with a cardboard box full of their office items. But the damage, in general, will have been limited to that company. With social technology, however, dangerous, money-driven people with such a personality disorder might bring down a nation with their lack of restraint. Could they destroy all of humanity? Yes. Will they feel any remorse? Not even the slightest bit.

Similarly, people who are addicted to power might find their way on the prospective enemies list. Addiction to wealth and power often go together, but they are not necessarily intertwined. I've worked at several universities, where people make little more than minimum wage, once you account for all the extra hours they put into their jobs. But, oh, do these people love to control others. They are nosy, and they are petty. They absolutely would *love* to be able to monitor the thoughts of their underlings, their colleagues, and their neighbors. Many of these people, unfortunately, have been dealt a bad hand of cards in life. They're miserable, and the small bit of pleasure they get comes from looking into the lives of others and calling them out for things they deem inappropriate. The brainphone will keep lots of people under their nit-picking control, and so they will push hard for its existence.

Another enemies list candidate might be people who just want to mess with other people. I'm thinking of trolls—the people who get on chat rooms (real-time discussions) or forums (asynchronous discussion threads) and always take the extreme, opposing view. They're obnoxious and disruptive. Imagine what they might do when communication among humans becomes hive-like, as instantaneous thoughts become shared by large groups. A couple of jerks really could throw a wrench in the machine. Since everyone would be connected all the time, everyone constantly would be dealing with these rogues.

Some not-so-obvious enemies? Here's one: humanity's inexplicable desire to eventually destroy itself. We keep building better and better nuclear weapons, knowing full well that one day we'll let them all loose. We keep working harder and harder towards creating robots that are fully self-aware, even as it becomes obvious that these beings could become exponentially smarter than us and take over the world.

Another human trait that's an enemy to us all: people's inherent desire to feel like they belong. Maslow accurately observed that, next to water, food, and the desire to "be fruitful, and multiply, and replenish the earth," a person's strongest need is to feel belonged by his or her community. Unfortunately, it is the need for belongingness that will serve the proliferation of brainphones well.

Yet another trait: people's desire to just go with the flow. Going along with change certainly makes for a less stressful life. We don't all handle change well, but we all *desire* to. The downside to this trait is that if a change is for the worst, we still desire to go along. When you were a kid, did your parents ever ask rhetorically, "Why did you do that just because your friends did it? If your friends all jumped off a cliff, would you join them?" I never understood this question. The obvious answer: "Of course I would." I think the natural human condition dictates that most of us wouldn't even ask what was going on. We would just follow the herd right over the edge. Maybe there's a person or two out there who would say, "Maybe we shouldn't jump off that cliff." Or, instead, write a book about why jumping off a cliff might be a bad idea. (See what I did there?)

Oh, one more prospective enemy you should consider. In this case, it's a prospective future enemy: the Artificial Intelligence that one day will be smarter than us.

So, consider these enemies, what they're up to, and why you should dispute them rather than go along with their brainphone idea.

You Should Consider Freeing Yourself from Technology Addiction

Determining the brainphone is a bad idea probably begins with determining that the smartphone is a bad idea. Yes, it's a necessary evil. But that still makes it an evil. It sure is tough to put down, even for people who are walking along cliffs (Each year, hundreds of hikers drop from cliffs and die while trying to take selfies.) or walking across a busy street (Each year, thousands of pedestrians are struck by cars and killed while texting.). As with many things in life, it is much easier to look back and determine something was bad than to be in the middle of it and think, "Yikes, this is bad." Kind of a combination of hindsight being 20/20 and knowing how to discern the forest from the trees. Too many clichés in one paragraph? I apologize. The point is, when people around you start getting brainphones, and when they start acting weird even though they don't see themselves acting weird, *you'll* be better able to see what's going on if you're free from smartphone addiction as it exists right now.

If you think, down the road, you might be one of the few people in this world to say "no thanks" to the brainphone, consider getting an early jump on the situation and become less addicted (if there is such a thing) to technology. Even if you manage to reduce your daily smartphone usage from, say, three hours down to two hours per day, you will find yourself improving both mentally and physically. Yes, cutting back on smartphone usage reduces stress and anxiety. But by reducing these feelings, you also reduce the amount of cortisol that is released into your body, which reduces the chance of chronic disease. Reducing time on your smartphone helps you sleep better and eat better (Buttimer, 2021).

In recent years, I've attended several in-person lectures on smartphone addiction, its consequences, and its cures. After the lecture, I've always approached the lecturer and asked what he or she does personally to cut back on smartphone usage. Their personal practices are much the same as one another's. And they are all pretty straightforward. (The practices are listed in Appendix B of this book.) Then again, the steps implemented in the Twelve-Step meetings at Alcoholics Anonymous or Narcotics Anonymous aren't all that complicated, either. But they work, and their effectiveness is scientifically evidence-based. The difficultly is in getting someone to do the steps.

People who advocate for breaking the habit of smartphone overuse suggest that you start by shutting off your smartphone at night. That is, don't keep reading your screen until you fall asleep and the phone drops out of your hand. Try putting your phone in another room for the last hour before you plan to shut your eyes, so that your brain can cool off from all that screen-induced stimulation. Do you use your smartphone as an alarm clock? Purchase an old-fashioned alarm clock (maybe even a super-old-fashioned one that winds up), so that you don't need the smartphone near you in order to wake up the next morning.

The experts I've spoken with suggest "grayscaling" your smartphone screen, so your brain is not constantly being mesmerized by the colors. (Full disclosure: This was the toughest

Consider shutting off your smartphone at night.

step for me to do. I like the colorful pictures in news stories. But then, that's the point, right?) Unfortunately, many of the newer models of smartphones no longer allow this option.

Most phones have a feature that allows you to track your daily screen time. Keep track. Just the act of tracking your time on your smartphone will have the immediate impact of reducing your time. Attempt to keep trimming it down. Shoot for less than three hours a day. Then less than two. If you start liking the quiet time, start shutting it off it at a regular time in the middle of the day for creativity or quiet prayer or reflection. (Oh, and remember: If you get the brainphone, your screen time jumps to twenty-four.)

Here's one more idea that I never heard from a lecturer but read in an article. (In fact, it was the first thing I did when I began cutting back on my smartphone time.) Sean Hargrave, who writes for the UK version of *Wired* magazine, rightly notes that smartphone technology "is similar to getting a puppy or having a baby. If you don't train it, it will soon train you." His suggestion is to shut off all of your phone's notifications and app prompts.

> Be ruthless. Do you really need to know if a tube line you don't use has a good service, do you care if a retailer you haven't shopped with for at least a year has a sale? Another option is to manage notifications as they come in. By holding down each pushed message a menu is opened up which allows further notifications to be blocked or made to at least arrive silently. (Hargrave, 2020)

He also suggests that you shut off social-media notifications from friends, especially ones who are sending you "nudges throughout the day to say someone liked [your] picture of breakfast on social media." If the app allows, turn off notifications from people who are not contacts or whom you do not follow. Do many people live for such notifications? Yes. But that's the point. They shouldn't.

If you follow these steps, you might decide, "Wow, maybe I really *was* addicted to that awful device, and it's nice to be freed from it." You also might decide, "I sure don't want that old addiction back, magnified a thousand times by a device in my skull."

You Should Delve into the Concepts of Individuality and Privacy

One of the earliest features of the upcoming brainphone is the capacity to communicate instantly with others, without speaking. The problem with such rapid transfer of information among large groups of people is that it likely will form human collectives, similar to the hives or flocks of animals. "Animals do move in groups governed by the collective," observes Earth science writer Alexandra Witze (2020). "Think of a flock of birds, a parade of ants, a school of fish—all are swarms like those envisioned by [science fiction novelist Frank] Schätzing, if not quite as murderous." Witze notes that scientists have figured out how information flows among the ants. It is as much about cooperative behavior and problem-solving bandwidth as it is communication, known as the Anternet. It's all about efficiency. But it has nothing to do with a single ant's individuality or its privacy.

Not to make a hyperbolic leap from ants to humans, but it is a question worth exploring. If information begins passing through brainphones rapidly and collectively, and if people, then, begin behaving as hives or flocks, do you want to be part of that set-up? If the answer is yes, I respect that. Some people live for being a part of a team. They recognize the desire, and they are comfortable with it. However, if you prefer a life that follows an individual path of your choosing, a hive's existence might sound horrific. Furthermore, if you value your privacy, being a part of The Hive—where everything about everyone, including you, passes through all brainphones—might be something that you decide is not a good idea.

This isn't the stuff of science fiction. I argue that smartphones and social media have brought us a lot closer to living a hivelike existence now than we might care to admit. Certainly, the concepts of individuality and diversity are celebrated. However, in practice, I don't see much of anything these days other than forced obedience. People are expected to conform to the group, in 100 different ways. They're pressured to say the same things and pretend to believe in the same things. I used to work as a dean for a university that specialized in national security. I once invited a discussion on whether or not separating immigrant children from their parents along the southern U.S. border was an effective deterrent to illegal entry into the United States. As a devout Christian, I had strong, personal reservations about the practice, but as a security professional and as an academic, I wanted to keep an open mind and hear what others had to say. I thought it would be a vibrant discussion. As it turned out, most of my colleagues in that meeting already had formed an opinion against the practice, and clearly it was an opinion that had been formed from strong emotions. They quickly shut down the discussion. Again, this was a university dedicated to studies in national security, with many graduates working among the different U.S. military, homeland security, and intelligence agencies. (Frame of reference reminder: I consider myself a moderate Republican voter.) But The Hive didn't want to talk about it that particular day, and I was forced to toe the line.

Ask yourself how comfortable you are these days sharing a deeply held view with someone. Is it a view based on your religious beliefs? Are you concerned about being shouted down? Are you afraid of being ridiculed, either online or in the real world, by a hyper-sensitive mob of self-designated, self-righteous, politically-correct Hivesters? Me, too. I think it's gotten way worse in recent years. People seem very agitated. At least, for now, we occasionally can shut off the electric static that seems to amplify the ailment. With brainphones implanted? Hmmm, not so much.

Ironically enough, a lot of Hive bullying these days is aimed at individuals who embrace the idea of a Hive but think it should have Hive standards. That is, if a person supports The Hive's official opinions, he or she is accepted. However, if the individual suggests that people within The Hive should display, say, mature behavior, he or she might be in trouble. If the individual suggests that people should embrace a sense of personal responsibility, he or she likely will be scorned. As you might have noticed, today's hives tend to love collective victimhood, and the concept of self-responsibility is so outrageous as to be viewed as, well, physically threatening. What bizarre quasi-reality are we left with? Individuality is fine, except when the individual opinion espouses responsibility, or when the individual opinion runs

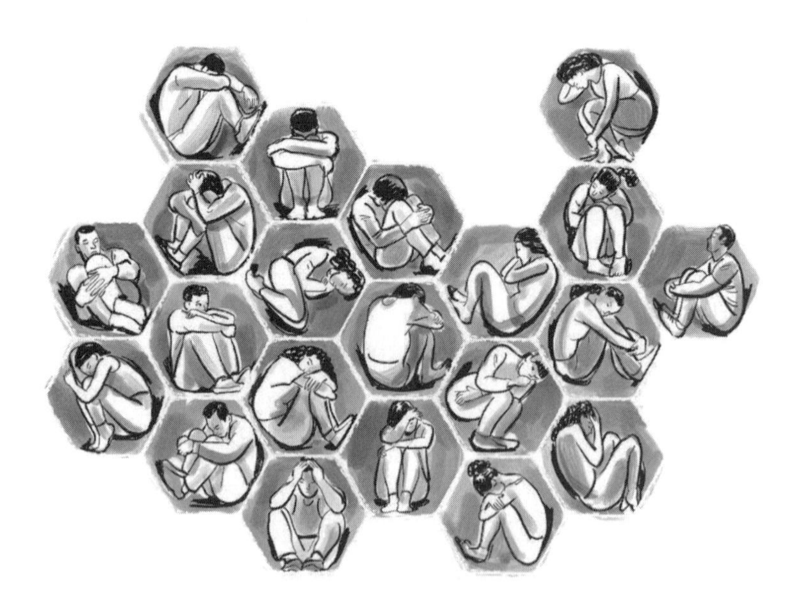

Do you really want to be part of a human hive?

contrary to The Hive's collective way of "woe is me" thinking. The Hive needs to feel safe, and this safety somehow gets transmitted by sameness of opinion. Confused? Me, too. Will the farce get much worse when we're all connected by thought? Yes, much worse—by a thousandfold. And the reality of the individual person will be lost forever within the spectacle.

If you read regularly, either at night before falling asleep or in the morning with your cup of coffee, check out some deep, historical readings concerning the First Amendment to the U.S. Constitution. Consider how the construct of freedom of expression has changed over time, pendulum-swinging towards the McCarthy persecution in the 1950s, then back towards liberty with major, supporting Supreme Court rulings in the1960s and 1970s, and then back again towards oppression in the 2010s, when the Court oddly declared that corporate propaganda and the political contributions of rich people were protected speech. With the muscling of social media and the hyped-up, instant backlash people get for speaking up nowadays (to include being fired from jobs that have nothing to do with the person's expressed opinions), it seems like the First Amendment has become almost as quaint as the Fourth Amendment. We believe in freedom of speech, as long as the speech shares the opinion that most other people are subscribing to at the moment. What is that shared opinion? It's hard to say—it's a rapidly moving target.

Enough about individuality. Let's move on to privacy. I was a little unnerved some years ago when I read about how many times the license plate of my car was being randomly scanned on any particular week, by either a police car's license plate reader, a plate scanner at a road intersection, or a road camera taking millisecond photos of plates as cars drove by. All this information is stored, and can easily be cross-referenced. I'm not crazy about the idea of being tracked, especially when there doesn't seem to be any regulation controlling the practice. Again, if a person isn't being sought after by the authorities, I argue that he or she should not be followed. If people paid attention to the U.S. Constitution these days (again, mostly the Fourth Amendment), the practice would not be allowed. With smartphones, the tracking is even worse. Your phone carrier likely has years of data on your precise daily locations, as referenced by their cell towers. Apps keep track of your location, unless you turn off the feature, so that they may send you location-related advertising. And, believe it or not, in many states, your boss is legally permitted to track your physical location, via your smartphone, even during your off hours.

However, license-plate tracking and smartphone tracking pale in comparison to how little privacy anyone has anymore as the result of the Internet. Someone can punch your name into a search engine or an app and find your address, your relatives, your relatives' addresses, your neighbors, and your neighbors' addresses. That person can find a photograph of your house, as well as when you purchased it and how much you paid for it. (Yes, I know anyone can find this information on file at the county courthouse. But it used to take a little time and effort to track down!) He can find out what social circles you hang in, and where you've traveled recently. That person can find out where you work, and where you've worked in the past. He can find out if you have a criminal record. If *that* person is a criminal, with a little bit of effort, he can find out where you keep your money

It might shock you how often a machine reads your license plate each week.
Photo by Obi Onyeador. Courtesy of the photographer via Unsplash.

and where you spend your money. With a little more effort, he can hack into your accounts. On the Internet, you're a deer grazing in the meadow.

Even more chilling is the profiling that companies perform on you these days using your data and online behavior. Companies gather an enormous amount of information on you and other consumers—information that is sold to other companies. They place "cookies" on your devices, used to identify you and record your searches. They cross-reference your behavior among different sites and apps you check out, including choices you make across different devices. Smart speakers in your home constantly listen to you and track your conversation patterns and inclinations. To its credit, the European Union is attempting to push back on some of these practices, regulating what companies are permitted to do with consumers' digital data, and getting consumers' permission to do so. The California Consumer Privacy Act is a similar piece of legislation, although its effectiveness at this writing is inconclusive.

Another Web-based threat to your privacy is your susceptibility to hacking. Some years ago, if you were to keep a written journal—with your deepest, most secret thoughts on the pages—in a lockbox in your home, an intruder would have to commit a couple of pretty serious crimes to read those pages. Nowadays, if you write down your thoughts on Cloud-based files, it is feasible they're being hacked and read by someone. If the camera on your device doesn't have a plastic slide-cover over it, someone might be watching you (even right now, if you're reading these pages via download!). If you're watching your child via a wireless baby monitor, a stranger might be watching her, too. If you're keeping track of who has been at your front door with a smart doorbell camera, someone else might be keeping track, too. And if they can hack this information, they can use it against you. After all, not all hackers are bored teenagers munching on chips in their parents' basement: some are professionals, and they make serious money by hacking your data, stealing your identity, and then using that identity to make purchases. In sum, due to the Web, much of our privacy has been whittled away to nothingness.

You Should Decide that Humans Should Be as They Already Are

Admittedly, some of the brain-tech devices being tested now have admirable goals. Helping a fully paralyzed person, who can't speak, to communicate with his caretakers, by implant, is compassionate. In fact, the devices that do these types of things, as currently designed, don't resemble smartphones at all, but are, instead, dozens of nodes, placed on the surface of the brain like a hairnet. Deep electrodes meant to stimulate the brain and rid a patient of seizures also strike me as having benevolent intentions behind them. The problem with the smartphone-like implants—the brainphones—is that they seem to be a solution in search of a problem. That is why, in my opinion, a few of these devices are being designed without a clear vision about what they're supposed to do. In the United States, there *has* to be some type of medical or mental disorder that the device will help remedy, or else the Food and Drug Administration will not allow the trials to happen. But make no mistake about it: Once people start receiving the implants for ostensibly medical reasons, then the companies will push for everyone to have one. The long-term goal is mass purchasing and mass implanting.

You almost can hear the commercial for the brainphone in your head, can't you? How would you like an upgrade? How would you like to be Human 2.0? Communicate without speaking! Understand foreign languages without having to learn them! Memorize huge amounts of information instantaneously! Visit new places, virtually, without ever leaving your living room! Play realistic, 3-D videos games, virtually, with others, the likes of which you've never seen before! Link yourself with Artificial Intelligence, and become smarter, faster, better! In other words—maybe it is time for an upgrade! Well, step right up, folks. Get your brainphone today!

Of course, as I've argued throughout this book, for every step forward, there are ten steps backward. The person fully wired to the Web and to others eventually will curl up into a ball and exist in the virtual world only. In many ways—physical and personality-wise—he will will become unrecognizable.

But, as you ponder whether or not the brainphone is a bad idea, perhaps there's a more basic question you should ask yourself: Do humans really *need* an upgrade? Or are we already the way we should be? If we forget who started any particular argument, might it not be by design? If the world is full of different languages and cultures, waiting to be learned over time, might that not be the spice of life? Are we meant to do high-level calculus in our heads without really understanding the concepts behind the math? Are we meant to live in virtual worlds, created by a gamer, when the real earth is so beautiful? I'll ask it again: Do we really need an upgrade (especially if it's likely to turn into a downgrade)?

As you're making these decisions for yourself, you might want to set a can of tough-love aside for when your children ask for a brainphone. In other words, maybe now is a good time to decide that it shouldn't be the kids who decide. We make lots of dumb choices as teenagers, and expectedly so: the human brain isn't fully developed until age twenty-five. (Auto insurance companies figured this condition out decades ago. That's why your insurance rates go down when you reach that age.) But some bad decisions cannot be reversed, which is why we, as adults, have to protect kids from making some of them. Remember, this is not science fiction: the earliest versions of the brainphone will be no more difficult to implant than to administer LASIK surgery. The Neuralink procedure involves a contraption (to be fully robotic at some point) that easily will be mass produced. In Chapter One, I offered a scenario where a teen argues forcefully with her mother just outside the Brainphone Store in the mall, shouting, "But mom, *all* the kids are getting brainphones!" Don't let your children bully you into letting them get the implant, and don't let them bully you into getting one for yourself.

Another question you might ask yourself: Is it necessary to fight Artificial Intelligence (AI) by joining it? Remember, that is one of the reasons Elon Musk has given for co-founding the company Neuralink. He believes we're going to design AI that is so smart, it will one day cause humans to be trivial (if not extinct). His solution: merge humans, via brain-computer interfacing, with the AI, so that we have a fighting chance at continued, meaningful existence. I couldn't disagree more. Transhumans will not be an improvement. The human-AI hybrid will, in my estimation, be a hideous creature. And, I suspect—because the

The stop-motion ED-209 model used by Industrial Light & Magic in the 1987 film *Robocop*. Photograph by Steve Jurvetson. License https://creativecommons.org/licenses/by/2.0/deed.en

AI will program them and brainwash them—transhumans are more likely to fight for their machine side than they are for their human side. In other words, transhumans will become allies with robots in the attempt to take over the world.

I take the position that once we realize brainphones, AI, and transhumanism are taking over the world, with a little bit of courage, we can still win as 100 percent humans. Transhumans and robots, I believe, will not be unbeatable. In fact, I think they will be rather pathetic. Remember the original *RoboCop* (1987) movie? Towards the end, a giant killer enforcement droid, called the ED-209, chases after Murphy. One problem for the droid: It never was programmed to climb or descend a flight of stairs. It comically falls down a few stairs in a failed attempt, crying like a baby as it rolls on its back and gets stuck there like a turtle.

There are a million ways to hack back at transhumans and robots, if a full-on war commences. On our side will be a legion of young, genius computer geeks who enjoy technology but don't want to be slaves to it. One possible, more-crude technique: We'll just pull the plug on 'em. I don't mean that metaphorically. I mean we'll literally steal their extension cords and watch them run out of power. Take that, ED-209!

CHAPTER 10: HOW MIGHT YOU REVOLT AGAINST THE PROPHECY?

The steps of a good man are ordered by the Lord,
And He delights in his way.
(Psalm 37: 23 NKJV)

You Should Draw the Line

Fortunately, the line between humanity and technology is still well defined. That line needs to be expressed through activism—perhaps political, perhaps religious—and it needs to be preserved. That is, once it becomes apparent why people should not physically merge with technology, then the campaign against such melding will gain traction. In my estimation, there needs to be a worldwide uprising against the brainphone.

As mentioned before, I'm not necessarily against all forms of brain hardware. A device, for example, that zaps the brain and helps control seizures—in and of itself—does not cross the line. Neither does, say, an artificial, digital spinal cord that helps a paralyzed person walk again. So, what *does* cross the line? This book ends the same way it began, by noting the differences between what is being promised now to what is likely to be delivered. For example, earlier possibilities regarding the Neuralink device (admittedly to draw hype and garner investors) included learning another language instantly; having brain diseases treated through brain reprogramming; and having our bad tendencies washed (brainwashed?) away. However, upon closer look, these planned devices disturbingly resemble smartphones, each one having a Bluetooth, a processor, and a wireless charger (with modems coming later on). Elon Musk himself noted that the device "is kind of like a Fitbit in your skull with tiny wires" ("Watch Elon Musk's," 2020). They are to be implanted quickly and painlessly in people's brains (by robots, no less!), en masse. Musk has said he wants to make the

Fortunately, at present, the man/machine line is still clearly drawn.

"product…affordable and reliable, and such that anyone who wants one can have one" ("Watch Elon Musk's," 2020). So, that's the line: technology that is mass produced, connected to us permanently, changes us in obvious and profound ways, and communicates with other technologies outside our bodies.

More importantly, I'm against the *soft*ware. In order to merge human thought with computer code, the thought itself must be converted to code. It is a scary form of reduction, when we reduce human thought to code. Where is the heart that goes into a thought? Where is the input from friends and family that helps mold that thought? What about the circumstances? The hormones? The good (or bad) intent? The nuance? None of these things exists in the code.

Furthermore, if human thought is reduced to code, it can be replicated, stored, and transmitted. It can be shared with everyone, all at once. Haven't we all had a thought that we were glad was just a thought? If everyone can know that thought, isn't it just as bad as acting on it? And, if that's the case, might not people be inclined to act on their bad thoughts, if they're going to be outed for thinking them anyway? A society where everyone knows everyone else's thoughts immediately seems like a society that is destined to crumble. Families and communities will deteriorate when bad thoughts can't be buried (like "burying the hatchet" and forgotten about).

But the reasoning doesn't have to be complicated or super enlightened. There's a basic motive to draw the line: Smartphones are harmful—mentally and physically. There is lots of scientific evidence suggesting so. And if you have a miniature smartphone implanted in your brain, the harm will be magnified. Forget the promises about brain advancement. Forget about instant knowledge. Forget about lifelike gaming. The brainphone eventually will devastate you and your family.

The benefit to drawing the line between humans and technology now is that you don't have to worry about getting to a point later on where you ask yourself, "Am I still me, or have I become something else?" Suppose your ten-gram brainphone is eventually replaced with a one-ounce brainphone, and then a one-pound brainphone, on up to, say, three pounds. Well, three pounds is the weight of the adult human brain. So, is there anything left? Are you still you? And, if you are not still you, at which weight did you stop being you? This incremental human-to-machine replacement can't happen if you stop at the beginning—saying no to that first, ten-gram brainphone.

If you find yourself addicted to your smartphone, and you really, really want it implanted in your brain—but you maturely recognize that it ultimately will bring harm to you—then you could go to a support group meeting. As I alluded to in the previous chapter, a Twelve-Step group similar to Alcoholics Anonymous might soon exist to convince people to cut back on their smartphones and to say no altogether to the brainphone. It also might include members who had received the implants, realized that their existence as humans was hitting rock bottom, had the implants removed, and are now dealing with serious withdrawal symptoms. Good things await them if they follow the Twelve Steps: eventually, they will go back to the wonderful lives they had had without the brainphone. What will the group be called? Brainphone Anonymous?

Speaking of drawing support from others, saying no to the brainphone is likely to work better if it is a family decision. That's not to say that the choice has to be unanimous: certainly, a parent's vote carries more weight than a child's vote. However, if there is at least a little bit of buy-in from all family members, it will be easy to draw from the strength of the family when life without brainphones becomes difficult. The family that prays together, stays (away from the brainphone) together. No reason to keep it in the immediate family. Rejecting the brainphone can be part of an extended-family commitment. It can involve neighborhoods and church groups. There can be clubs specifically geared toward rejecting the brainphone. Brainphone Banishment Brothers, Post #77. Meetings at 7pm on the first Tuesday of each month. Spouses welcome to attend. Beer and bingo follow the meeting.

You Should Push Back as a Voter

If it were up to me, there would be no such thing as a career politician. A person who wanted to make a difference in government would run for office, serve one term, and then go back to his or her life and regular occupation. Imagine how little influence corporations and rich donors would have if their money wasn't being poured into the continuous re-election fundraising of lifelong politicians. However, if there is one good thing about career politicians, it's that their single focus on getting re-elected keeps them in tune with what their constituents (especially their loudest constituents) are concerned about. Write to your U.S. representative, your two U.S. senators, and the president. (There are sample letters in Appendix A of this book.) Let them know that you are a one-issue voter, and the issue is the brainphone. That is, you are opposed to brain-tech implants for mass production and insertion in human skulls, and that you support federal legislation banning it. You will be voting for politicians who favor outlawing the brainphone, and you will be voting against politicians who support the brainphone (particularly those who are accepting re-election funding from brainphone manufacturers). Oh, and if you do write to a politician, consider putting the letter on paper and mailing it. More on why later on.

Not all campaign money has to come from big donors.

Write to your U.S. representative, your U.S senators, and the president, asking for a brainphone ban. Photograph by Andy Feliciotti. Courtesy of the photographer via Unsplash.

Consider donating a small amount to whichever politicians, political parties, and political action committees are being the most vocal against the brainphone. If you are supporting such politicians, let them know that you are doing so specifically because of the stand they are taking against the brainphone. Does one dollar matter to a politician? No. Does one dollar each from ten-thousand donors of like mind matter? You bet.

Along with politicians, write to corporations and ask them to take a stand against brain-tech implants. Not too many years ago, corporations used to avoid taking political positions the same way one might avoid being around the plague. But then politics became extreme, and people began calling out companies via social media. They also began arranging boycotts that way. Some were effective in reducing sales with the goal of bringing about corporate change. Most of the proposed boycotts didn't do much more sales-wise than scare companies into being proactive. Either way, the social media campaigns worked. These days, it is relatively common for companies to take political and ethical stands, and to scale back their operations in states with oppressive or discriminatory laws. This new way of projecting a corporate image might not be a bad thing. After all, if a huge corporation knows it's going to be put on the spot, anyway, it might as well channel its vast resources into taking a position and having a hand in whatever types of societal changes are being promoted. I suspect that, if enough companies receive enough letters and online attention over their stands regarding the brainphone, they might join the resistance. If a few of these companies are part of the brainphone's manufacturing supply chain, and their actions stall production/distribution of the devices, your communication with them ultimately could have a real impact.

Speaking of social media, by all means use it. Yes, I know I've spent a sizeable portion of this book denouncing social media. And for good reason. However, I think we all would agree that they're not going anywhere. And history (as well as religious lore) is full of instances where some malevolent thing was repurposed for something good. If social media can be used to start conversations and convince at least some people not to get brainphones, then maybe some of social media's more horrible aspects can be forgiven. It's doubtful, but maybe.

If you'd like to be an activist against the brainphone, give presentations. Any planned event at a church, community center, or school/university auditorium will do. Bookstores often host such presentations. Did only five people show up? No worries. I've been there, too. It's five minds you might have changed. Five lives that will flourish after the brainphone commandeers and destroys the lives of everyone else.

If you're going to push back against the brainphone by convincing others, stay knowledgeable about the issue. Things are changing rapidly, and there are new iterations of brain-machine interfaces being introduced almost every day. Of course, reading this amazing book already has made you very educated on the topic. Regularly type "brain computer interface technology news" into Google or Bing and check out the latest top stories. As mentioned before, don't get all of the news from one or two sources. Mix it up. And be aware of the backgrounds (and intentions) of the people who own the news sources you're watching and reading.

If you're going to advocate against the brainphone, try as best as you can to walk the talk. After all, it'll be tough to rail against the device if you've had one implanted! If you're going to argue that the brainphone is a smartphone with amplified, harmful effects, set a good example by cutting back on your smartphone use as well. Seek out people who have the same passion. Enjoy their company and their *example*. Find out what they are doing to live a brainphone-free existence. Select those aspects of their lifestyle that suit you and incorporate them into your reality.

Engage in respectful conversation with people who disagree with your anti-implant stance. Listen to what they have to say. Engage in active listening—lean forward, nod your head, mirror back to them what they're saying. Show you're interested. You'd be surprised how much influence you can have over other people simply by taking the time to engage them in their opinions. It's called the Rogerian Argument. You should look it up some time.

Finally, stay peaceful. If there's no peaceful way to keep corporations from manufacturing their horrifying brainphones, then just let them manufacture them. There will be other, peaceful ways to push back later on. In the early 1970s, Berkeley math professor Ted Kaczynski left society, moved into the woods, and wrote passionately about how technology was causing psychological damage to humans and destabilizing humankind. Had he stayed peaceful with his passion, he might have had an impact. Instead—in ghastly fashion—he chose to anonymously mail bombs to people connected in various ways to industry. He killed three of them and injured dozens of others over a 17-year period, becoming the infamous Unabomber. His brother, David, reading Kaczynski's anonymously published manifesto, recognized the writing style and contacted the FBI. Kaczynski, at 80, is still in prison. His violence not only had zero impact on his anti-technology aims, but his actions destroyed and ruined lives. Rather than embracing the spirit of his back-to-nature lifestyle, he instead chose to be a killer, a terrorist. (Granted, he might not have had any choice at all: he might simply have been crazy.) If you want to push back against oppressive technology as a voter and as an activist, you only will influence outcomes for the long-term if you do so peacefully.

You Should Push Back as a Consumer

In the interest of delving into people's thoughts and keeping huge files on them, corporations and governments will endeavor to persuade people to get brainphones. As mentioned earlier, many people will take very little coaxing and will buy into the wonders of the technology. Others will hold out, until some sort of contrived emergency comes along and brainphones are offered up as the solution. Eventually, when 90 percent of the human population has implants, the rest likely will be made to do so.

However, I argue that the forcing of brainphones on people cannot come before the consumerism. That is, for the brainphone to be fashioned in a way that it takes over society, a lot of people first must *want* to have it before the remainder are *forced* to have it. It is the consumer dollars in advanced nations that will fuel the perfection of the brainphone and the network it runs on. Conversely, if there are no consumers, the project might starve—cut at the roots, so to speak.

Admittedly, it is possible that just the gamers alone will fund the manufacturing of brainphones to the point that it can forced on all others. If statistics are to be believed, there are more gamers in the United States right now than there are people who don't play video games. But I'm not sure that the passion for permanent implants to enhance the gaming runs throughout the gaming community. After all, lots of gamers are also hackers, and a percentage of hackers appreciate the concept of privacy. They also like the ability to remain anonymous. Privacy and anonymity quickly would become things of the past if brainphones were to become universally used. And, of course, there is still a sizeable portion of society that doesn't game at all. Are they all boomers? Ha! Maybe. But boomer money is big consumer money, and so boomers impact the marketplace.

You also can leverage your sway as a consumer by refusing to purchase smartphones from companies that are selling brainphones. You can refuse to purchase wireless network service from companies that are providing such service for brainphones. Write to companies that refuse to deal with the brainphone. Let them know that you're proud of them and that you, your family, and your friends will continue to support them as consumers because of the ethical position they are taking.

Behind closed doors, someone somewhere has decided that not supporting the brainphone might, in the long run, be financially good for the company. Your letter will make them feel better about that decision.

Consider purchasing and endorsing products that enhance, rather than hinder, personal privacy. Cybersecurity author and consultant Kevin Mitnick wrote a great book not too long ago, entitled *The Art of Invisibility: The World's Most Famous Hacker Teaches You How to Be Safe in the Age of Big Brother and Big Data* (written with Robert Vamosi; Back Bay Books, 2017). He reveals all kinds of ways corporations and governments are tracking you, as well as the types of gadgets that can protect you from being tracked. For example, if you plan to attend a political rally, but you don't want the police adding you to their facial-recognition database, you can purchase a *privacy visor*. Mitnick explains the device:

> The eyeglasses [from the National Institute of Informatics in Japan], which sell for around $240, produce light visible only to cameras. The photosensitive light is emitted around the

Purchase products that enhance personal privacy. Photograph by Chris Yang. Courtesy of the photographer via Unsplash.

> eyes to thwart facial recognition systems. According to early testers, the glasses are successful 90 percent of the time. The only caveat appears to be that they are not suitable for driving or cycling. (Mitnick & Vamosi, 2017)

Pretty nifty, right? On the day-to-day-life side, consider purchasing products that don't automatically connect to the Internet of Things, such as doorbells and baby monitors. Consider conducting your social-media life on platforms that aren't as quick to sell information about you and your interests to the highest-bidding advertisers. Again, let them know that you're using their service because you appreciate their commitment to at least some baseline of privacy.

Once a few autocratic nations have mandated the brainphone, consider purchasing products that are not imported from those nations. The only thing worse than supporting slave labor is supporting human labor that is being hypnotized and controlled by brainphones. Admittedly, limiting your purchases in this manner is a bit of a tall order: it's difficult to track down where all the components of certain items are manufactured. Plus, it's tough to find—or afford—products that are manufactured entirely in, say, the United States. But do what you can. An oppressive nation, forcing its citizens to merge with machines, should be ostracized as much as possible.

You Should Push Back as a Prepper

In case you're unfamiliar with the term, a *prepper* is someone who is preparing for the Zombie Apocalypse. Okay, not really. But he or she *is* preparing for any one of several other types of apocalypses, including nuclear attacks, pandemics, and general, rapid breakdowns in society. Preppers—or survivalists—come in different forms, and they practice their craft in varying degrees. I dare say prepping probably seems silly to many of us—or at least it did until the COVID-19 pandemic came along. All of a sudden, having a several-months' supply of toilet paper didn't seem all that extreme. I find it particularly interesting that there are survivalism enthusiasts on both the left side and the right side of the political spectrum. Observes survivalist researcher Bradley Garrett, "Prepping is, at its heart, a kind of activism, a bulwark against the false promises of capitalism, of the idea of endless growth and the perpetual availability of resources. [And, on the right side,] this kind of preparation folds so well into conservative narratives distrustful of big government, experts, and elites. The right-wing commentators Sean Hannity and Alex Jones advocate doomsday prepping to their audiences, and market products to meet those needs" (Garrett, 2020).

I'm not suggesting that you build a fortified bunker to keep out the brainphone drones who used to be humans. (Well, at least not yet, I'm not.) However, I am suggesting that, one day, your boss might tell you that, if you want to keep working for his company, you're going to have to get a brainphone. If you don't want a brainphone, then it will be much easier to say no, get fired, and continue providing for your family if you have some cash and supplies on hand to help you keep going for a while. Speaking of cash, you might be told by your bank that you can't keep your money there anymore unless you have a brainphone to access the electronic currency. Again, having some old-fashioned paper money hidden in your walls might be a good idea if you want to stay solvent.

When the brainphone network is turned on and people begin communicating and doing business online, there will be tremendous pressure placed on people who prefer not to have the implant. There will be a list, and people avoiding the brainphone will be on it. However, you don't necessarily have to make your way on that list. I argue that the list will be contrived by tracking people online who do not have the brain-machine interfaces. In other words, the powers that be will surf the Web, figure out who is using an old-time smartphone or laptop, and put the squeeze on them. So, if you don't want to be on the list, keeping a low electronic profile might help. Again, I refer to Kevin Mitnick's great book, *The Art of Invisibility*. He suggests that keeping an anonymous online profile essentially means creating an anonymous online identity. His steps include purchasing a separate laptop for when you want to remain anonymous; not using a wireless router tied to your home; using gift cards to pay for wireless network service; and so on. Mitnick says that, given enough time, people are likely to figure out your identity. The key is to make it very difficult for them. "All you are really doing by trying to make yourself anonymous," he writes, "is putting up so many obstacles that an attacker will give up and move on to another target" (Mitnick & Vamosi, 2017).

Once the brainphone, its network, and its overseers begin running everything, if you don't want to be a part of The Hive, you'll likely have to go into serious survivalist mode. Will you have to exist in a cabin in the deep woods? Or perhaps a bomb shelter built underground in your backyard? Probably not. In fact, you might have the run of the land to yourself: most people with brainphones will be curled up in dark rooms, living in their virtual worlds while the real world passes them by. Cities might look quite dead, even on weekdays.

But, to the extent that you might choose to leave Brainphone Society, you should do a few Prepper types of things to get ready. As I mentioned before, having some cash on hand is a good idea. So is having other items that might be used in place of currency, like toilet paper and cigarettes. (My wife refuses to allow me to keep cigarettes in the house for bartering. We don't smoke, and I see her point that it might promote smoking among people who otherwise could have used the Brainphone Apocalypse as an opportunity to quit.) Gold seems to hold its value. I like the advertisements for those 50-gram gold bars that are scored (dented with lines), so they can be pulled apart and used as one-gram bars. Kind of like breaking a large bill into smaller bills. Make sure you research purchasing gold before you do so. Check out the Better Business Bureau reviews, and read articles telling of people, particularly seniors, who have been ripped off. For every reputable gold dealer, there is at least one that regularly sells overpriced items, particularly standard coins portrayed as rare or collectible.

Speaking of money, there is something to be said for being as debt free as possible. I like the type of financial strategies put out by experts like Dave Ramsey. Their philosophy is that debt is basically slavery. I concur. Is there a way some corporations, in the future, could somehow leverage your debt to them to force you into getting a brainphone? Sure. Slavery is slavery, no matter its form.

Look up survival lists of things you should carry to evacuate an area, such as for a hurricane or wild fire. Here's a very good list from the U.S. Federal Emergency Management Agency (FEMA): https://www.fema.gov/press-release/20210318/proper-emergency-kit-essential-hurricane-preparedness. My suggestion: Triple the amounts. You should be able to live on your own resources for a couple of weeks.

No need to be extreme. No need to spend exorbitant amounts of money preparing for the end of civilization as we know it. No need to become obsessed with the Brainphone Apocalypse. After all, life is for living—not planning for the end. But if you and your family are ready for a long-term emergency, you're likely also ready to say "no" when the brainphone enthusiasts ask, "Are you ready for yours?" My friend Gary McCabe further notes that learning how to hunt, fish, and garden offers two survivalist benefits. "First, it prepares you for living without depending on tomorrow's technology," he suggests. "Second and more important, it gets you into the outdoors, and away from today's technology, which has a restorative effect."

One more thought about prepping. Long before everything goes to heck and humanity falls before the brainphone, there is a

Have a talk with your family about brainphones. Photograph by Allen Taylor. Courtesy of the photographer via Unsplash.

messaging side to being a prepper. If, say, millions of people begin stocking up, shutting off devices, and tuning out of the brain-tech economy—in essence, creating their own separate economy—corporations and the government might back off from brainphone mandates, at least for a while. In other words, if lots of people say, "We're not becoming part of the prophecy," then perhaps it might slow the prophecy down.

You Should Push Back as a Person of Faith

My friend, church pastor Gary McCabe wonders if turning away from the brainphone might ultimately be looked at as a religious statement. He asks rhetorically: "Are we going to end up being the new Amish?"

What are the differences among the various religions? I don't concern myself with them. In fact, I step back, and I see way more commonalities than I do differences. I'll let the theologians work out the finer points. And I'll let the extremists fight over which faith is the most preferred by God. Seems like a silly (and rather ironic) fight to me. For this text, I refer you to the generalities. Most religions put forth a set of beliefs and practices, or traditions. They worship a supreme being or beings. They possess old, sacred readings, often detailing how one goes about becoming the best and most enlightened person possible. I argue that, no matter which one of the major religions you practice, if you practice any of them all, then you should be pushing back against the brainphone as a person of faith.

Nearly all religions ask, "What is The Golden Rule?" The answer, of course, is that we should treat others as we would have them treat us. Furthermore, we should be concerned with our neighbors and how they are faring in life. I'm reminded of the 2002 movie *Equilibrium*, with Christian Bale. In the movie, everyone in society is forced to take a drug that suppresses their emotions. The idea is to keep wars from happening. Spoiler alert: If you're thinking of watching this movie, skip over the rest of this paragraph. Anyway, at the end of the movie, we discover that the people running society do not take the drug, and their emotions (and tendencies toward violence) are very much intact. People of faith should be opposed to the brainphone because, while the minions are being monitored and controlled, the very rich and powerful will be removed from it all. I'm not (necessarily) suggesting that the overseers won't have brainphones. They might (especially if they are robots or vast networks of Artificial Intelligence). But if they do, they most certainly will be removed from the standard, mass network being manipulated. Does this condition violate The Golden Rule? You bet.

Most religions consider the question, "What is the problem with people?" Christians, for example, believe that people are essentially born with sin and there is no way they can save themselves. It is the death and resurrection of Jesus that saves them. It doesn't stop with original sin. *The Holy Bible* itemizes, in detail and through stories, the many bad characteristics we humans have. No need to make a list. If you are a member of any good-sized family in this world, you've seen all the attributes among your various relatives, anyway. (Check out a few more by looking in the mirror!) In fact, this book has covered some of the bad attributes of people, including their vulnerability to addiction, their self-centeredness, their tendency to blur reality

Be wary of leaders who say they're getting a brainphone right after you do.
Photograph by Toa Heftiba. Courtesy of the photographer via Unsplash.

and the realities of others, and an overriding desire to feel accepted. These are all conditions that will be magnified by the brainphone. Therefore, people of faith should be inclined to say, “Here’s what the problems with people are, and we don’t need the brainphone making them a zillion times worse.”

Once a religion identifies what it deems wrong with people, it then asks, “What is needed to remedy this problem?” Not only will the brainphone make human conditions worse; I argue that it is unlikely to fix whatever ailment it is seeking to remedy. As I’ve mentioned before, the brainphone seems to be a solution in search of a problem. Once you get past the investors, the PowerPoint slides, the amazing algorithms, and the genius minds behind the device, all you really have is a bunch of tech geeks saying, “Hey, let’s put this gizmo in people’s heads and see what it can do!” Doesn’t match any scripture I’ve ever heard of. And I certainly don’t think the brainphone will improve the goodness in people to the point that they get into Heaven. In fact, I have many, many questions concerning the free will of humankind, and what impact the removal of free will will have on people’s

Will a person who’s mostly machine ever be allowed to die?

souls. But that's another essay for another time.

Finally, many religions claim to answer the question, "What happens after we die?" Believing what happens to us after our time as mortals is the centerpiece of many faiths. Unfortunately, it is in this area where the believer has much to worry about. Say, for example, that your brainphone carries all your thoughts, all your experiences, and—as close as the algorithm can match them—all your emotions. Then you die. Can your brainphone then be moved to the brain of another person and have you continue to live? Or can your essence or persona be placed in a robot, and you become who you are, only in a robot's body? What about your soul? And what if you live in perpetuity, jumping from human to human or robot to robot? Will you ever be truly dead? If not, does that mean you'll never see Heaven? Life on Earth is tough—does anyone really want to live forever? I vote no for myself. These questions challenge the very reasons why many religions believe we're on Earth and what awaits us in other realms. The brainphone throws a wrench into the machine of any religious faith, except, perhaps, the worship of money and materialism. And so, people of faith should push back and reject the device.

You Should Recognize that One Person Can Make a Difference

Some years ago, one of the largest corporations in the world decided they were going to turn the dead-end street I lived on into a truck entrance at one of their properties. The entrance wasn't needed: there were plenty of ways into the processing facility located on that piece of land. This new entrance would have meant huge, flatbed trucks carrying bulldozers going up and down my narrow street all day and night. Kids who played on the dead-end street would have been in danger. It would have completely changed the dynamic of that quiet neighborhood.

As it turned out, the corporation never got permission from the township to build that entrance. So, when the trees started coming down and it became obvious what was happening, my neighbors, my wife, and I called a township zoning inspector we knew, and he stopped the workers who were clearing out the trees. Later that week, the township told the corporation they would have to go before the zoning commission, at their monthly meeting, for permission to put in that entrance.

With about 30 dollars' worth of copies announcing the meeting, my neighbors, family, and I went around the neighborhood and asked people to show up, expressing their opposition to that truck entrance. The night of the meeting, the room was packed with my neighbors—nearly 100 of them! The corporation had about half a dozen engineers and lawyers on hand. I later found out they had spent 60,000 dollars on legal fees and engineering studies to go before the zoning commission. The commission voted down that entrance. Thirty dollars beat 60 thousand. I no longer live in that neighborhood, but the last time I checked, my old street was still a dead-end street, and the area was still thriving with young families.

Just a couple of years ago, my grown daughter, a paramedic, was complaining about a bad intersection near my new home. There had been many car crashes at that intersection, and my daughter, while on ambulance duty, had pulled several severely injured people from their totaled vehicles. The problem was easy to discern: A small but busy access road crossed through a major highway. It wasn't quite busy enough for a traffic light, but it needed a STOP sign. Instead, there was a YIELD sign there, and people really weren't paying attention to it. And so, drivers were drifting onto a busy highway and getting T-boned by fast-moving cars and trucks. It was a mess, but no one seemed to be doing anything about it.

One day, I parked nearby, walked to the intersection, and set up there with a lawn chair and a digital camera. I took a few pictures, including photos of some near misses during the short time I was there. I called the state's regional Department of Transportation office, got the name of one of their "decision-makers," and wrote the gentleman a friendly but urgent letter, with colorful photos to get his attention. About a week later, he called me to discuss the letter. About a month later, state workers took down the YIELD sign and put up a STOP sign. My daughter was floored that, after all those accidents, it was one person's letter to one government administrator that made all the difference.

I believe in Heaven, and I believe that, one day, I will stand before God in judgment. I'm the first to admit to anyone these days that I am a miserable, failed human being. I'm not worthy to stand before a flea-bitten, old, mongrel, much less the Creator of the heavens and the earth. But if, by chance, I'm saved and fortunate enough to enter the Gates, I'm guessing it will be, in part, because I saved a few kids from being run over by trucks in my old neighborhood and saved some motorists from car crashes in my new one.

The thing is, one person *can* make a difference. In fact, I would say that, these days, one person can *really* make a difference. As you might guess, I'm not a fan of social media. But sometimes bad things can be spun around for something good. Activism against brainphones, I believe, is something good. And since social media allows for one person, in the middle of nowhere and with no other type of platform, to gain immense popularity, perhaps a few current and would-be Influencers might take up this good cause.

On the other hand, a problem with social media activism is that it often stops at posting or tweeting some watered-down opinion, and then forgetting about it. But there's a silver lining there, as well: The people who do a little bit more than post-and-forget are likely to have more sway than they might have had 20 years ago. For example, years ago, I suspect that politicians received a lot of paper mail, regarding citizens' concerns on a variety of issues. Nowadays, people (or machines) send them an email, to which the politician's office (or machine) automatically sends a courtesy response. Political advisors might keep tabs on what topics and opinions are trending, but it's all very fleeting. A paper (typed or handwritten) letter, on the other hand, is likely a rarity—which means that people remember them. These old-school, tangible items have real impact, like the paper letter I sent to the state Department of Transportation, discussed above.

I also might add that, for those of you who are religiously inclined, your church's pastor might have some very good ideas on how to push back against the brainphone, from the perspective of activism. Pastors are very good about going out into the community and impacting the opinions of people. Examples might be brainphone-awareness cookouts, brainphone-awareness days of community volunteering, podcasts with politicians, and brainphone-related Bible reading and discussion—all things that can be planned and hosted by one enthusiastic person.

The brainphone is here. It's about to mesmerize humankind—and then go about destroying it. I hope there will be at least a small contingency of people in this world who choose not to have one and to convince others to join them in the resistance. When the world at large, packed with AI/human hybrids, flames out in one, giant Armageddon-like computer crash, it will be this small contingency that's left to rebuild civilization and regrow the race. Maybe, next time around, we could cool it with the digital devices!

CONCLUSION: WHY WE SHOULD FIGHT THE FUTURE

For the time is coming when people will not endure sound teaching, but, having itching ears, they will accumulate for themselves teachers to suit their own passions, and will turn away from listening to the truth and wander off into myths. (2 Timothy 4: 3-4 ESV)

As I was wrapping up writing the last couple of pages of this book, a few brainphone-related stories hit the news cycle all at once. Coincidence? Maybe. Omen? Ha! That gets my vote.

First, the American Institute of Physics published a report on brain-computer devices in their scholarly journal *APL Bioengineering*. The paper is more like a warning. It suggests that, with technology's powerful interaction with human brains, there will be profound social and ethical consequences. Some of the researchers' concerns: 1) Brain-computer interface (BCI) technology might intertwine with people in such a way that they are no longer able to make their own, autonomous decisions; 2) Corporations will regularly infringe on the privacy of people's deepest thoughts by gathering and reselling their brain data; 3) People might become super-addicted to BCI, with horrific physical impact, similar to that of the opioid epidemic in the United States (Portillo-Lara et al., 2021).

Perhaps most horrifying of all, write the authors, is the possibility that people under the spell of this type of technology won't even *realize* that they are no longer making their own choices. They note:

> [O]wing to the lack of proprioception, the human brain is unable to acknowledge the influence of an external device on itself, which could potentially compromise autonomy and self-agency. Because of this, users may be liable to mistakenly perceive ownership over behavioral outputs that

> are generated by the BCI, as well as incorrectly attribute causation to it...In turn, this diminished agency and undermined sense of self could even prevent users from being considered as autonomous agents with decisional capacity. Therefore, the design of mass marketed BCIs should aim to prevent impingements on user autonomy, as well as minimize the risk of dependency and impaired self-perception. (Portillo-Lara et al., 2021)

Scary stuff. The authors, as researchers tend to do in writing these types of analyses, call for better oversight. They suggest more transparency in design. They also sound the alarm for better—and more ethical—big-picture planning moving forward. Fat chance.

The second brainphone-related news story was about how brainphones might reach all parts of the brain. Rather than neurologists discovering some type of brain portal—which I predict in this book—this news report suggests that newly invented, super-small transistors might be peppered throughout the brain and act as boosters for the main implant. Scientists from three universities and the semiconductor company Qualcomm gave a demonstration of the rice-sized sensors, called *neurograins*. The presentation included a rat whose brain's cortex had been covered, via surgery, with 48 of the tiny transistors. The scientists on hand claimed that someday a human brain could hold thousands of the sensors (Coxworth, 2021). A photo in the story shows the magnified tip of a human finger with nine of these neurograins setting upon it. If they were any smaller, they would be resting in between the ridges of the fingerprint. It seems to me that similar, man-made seeds could be placed in a human's blood stream via a syringe in the arm and make their way to the heart or the brain or wherever they might be needed. This article, in particular, concerns me. It is one thing to say, "Hey, get this brainphone out of my head!" It is quite another thing to say, "Hey, get this brainphone and these two-thousand mini-sensors out of my head!" The more the brainphone morphs into something that is not fully self-contained, the more difficult it will be to stop or reverse the horror of its grand-scale

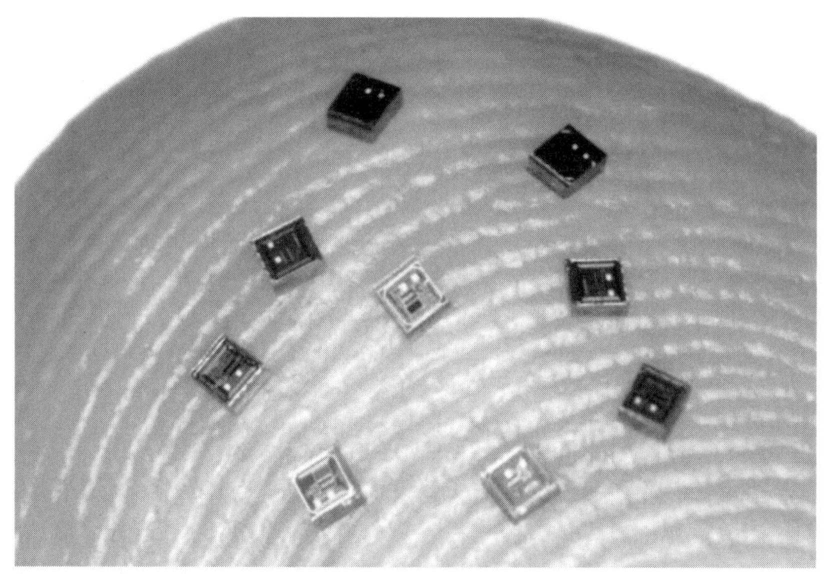

Neurograins—tiny transistors to be peppered in the brain and act as boosters for the main implant. Photograph by Jihun Lee. Courtesy of Jihun Lee.

implementation.

The third news item to hit just as I was finishing this book concerns Max Hodak, who co-founded Neuralink Corporation with Elon Musk in 2016. Hodak quietly left Neuralink in April 2021 without public explanation. This news story reveals why: He is starting up a competing brain-tech company! The company is called Science Corp. Hodak is hiring, and he has secured nearly $50 million in initial funding (Spichak, 2021). Hodak's specialty is research involving the reading and encoding of brain activity. This book presumes that Neuralink is likely to be the first company on the market with the brainphone. But with Hodak running a new, competing company with wheelbarrows of start-up money, things just got a lot more interesting.

Things are moving fast, that's for sure. Aside from perhaps another breakthrough or two needed in miniaturization and brain mapping, the brainphone is basically ready for mass production and mass marketing. However, it is not too late for us to push back. At this writing, no one publicly has a fully functional brainphone implanted in his or her skull—texting, Web surfing, and communicating (sans talking) with it. (Admittedly, these things probably already are happening in a secret, government lab somewhere.) Once a few young, telegenic people (social media influencers?) are in the news showing off their new brainphones and talking about how cool the devices are and all the neat things they can do with them, we will have reached a bit of a tipping point.

We should be writing to our governments and to corporations asking them to stop the experiments. We should be active in our refusal to purchase brainphones or to have them implanted if offered to us. We should be envisioning technology-minimalistic lifestyles if, down the road, other people have brainphones and we refuse to participate.

The reasons for resisting the brainphone strike me as pretty clear. Having something inserted in your brain is inherently unsafe. Having human thoughts reduced to code, and then watching as that data is sold on the open market, is disreputable and a clear violation of people's inalienable rights. It is, well, sinful. And, finally, the brainphone produces a hideous creature—the human/AI hybrid.

Let's reject brain-tech implants for mass production and mass insertion in human skulls. Let's push back. Let's fight the future.

ABOUT THE AUTHOR

Dr. Scott Snair is a data analyst and author in Washington, DC. He has nearly 20 years of experience working with and teaching about technology, analysis, research, and writing. His books have been published in 10 languages throughout the world. Scott maintains his certification as a data-privacy engineer with the IT professional association ISACA. He holds a Ph.D. in higher education from Seton Hall University and a B.S. in management from West Point, where he was president of his class. He chairs several doctoral dissertation committees at Grand Canyon University. Scott has testified before the U.S. Senate on military veteran issues, and he has appeared on CNBC with Jim Cramer.

BanTheBrainphone.com

REFERENCES

Ainsley, J., & Collier, K. (2021). Colonial Pipeline paid ransomware hackers $5 million, U.S. official says. *NBC News*, May 13, 2021. Article. https://www.nbcnews.com/tech/security/colonial-pipeline-paid-ransomware-hackers-5-million-u-s-official-n1267286

Albright, J. M. (2019). *Left to their own devices: How digital natives are reshaping the American Dream*. Amherst, NY: Prometheus.

American Academy of Pediatrics. (2016). Why to limit your child's media use. *AAP* website. Year available, but not date. https://www.healthychildren.org/English/family-life/Media/Pages/The-Benefits-of-Limiting-TV.aspx

American Psychiatric Association. (2013). *Diagnosis and statistical manual of mental health disorders* (*5th ed.*). Washington, DC: American Psychiatric Association Publishing.

American Psychiatric Association. (2017). What is addiction? *APA* website. Year available, but not date. https://www.psychiatry.org/patients-families/addiction/what-is-addiction

"'Anonymous' browsing data can be easily exposed, researchers reveal." (2017). *The Guardian*, August 1, 2017. https://www.theguardian.com/technology/2017/aug/01/data-browsing-habits-brokers

Appleby, L. (2021). Gabe Newell says brain-computer interface tech will allow video games far beyond what human "meat peripherals" can comprehend. *One News New Zealand*, January 24, 2021. Interview. https://www.tvnz.co.nz/one-news/new-zealand/gabe-newell-says-brain-computer-interface-tech-allow-video-games-far-beyond-human-meat-peripherals-can-comprehend

Bak-Coleman, J. et al. (2021) Stewardship of global collective behavior. *PNAS*, July 6, 2021, 118 (27). https://www.pnas.org/content/118/27/e2025764118

Becker, J. (2016). The cost of convenience. *Becoming Minimalist*. Website. Year available, but not date. https://www.becomingminimalist.com/the-cost-of-convenience/

Buford, B. (2016). Detoxing from addiction to success. *The Business Journals*, September 15, 2016. https://www.bizjournals.com/bizjournals/how-to/growth-strategies/2016/09/detoxing-from-the-addiction-to-success.html

Buttimer, D. (2021). Seven reasons to break your smartphone addiction. *Piedmont Healthcare*. Website. Accessed on July 24, 2021. https://www.piedmont.org/living-better/does-your-smartphone-cause-anxiety

Carman, A. (2020). Tinder made $1.2 billion last year off people who can't stop swiping. *The Verge*, February 4, 2020. https://www.theverge.com/2020/2/4/21123057/tinder-1-billion-dollars-match-group-revenue-earnings

Cave, S. (2016). There's no such thing as free will, but we're better off believing in it anyway. *The Atlantic*, June 2016. https://www.theatlantic.com/magazine/archive/2016/06/theres-no-such-thing-as-free-will/480750/

Coxworth, B. (2021). Tiny implants could dramatically improve brain-computer interfaces. *New Atlas*, August 12, 2021. Article. https://newatlas.com/medical/neurograins-brain-computer-interface/

"Cutting." (2021). *Johns Hopkins All Children's Hospital Health Library Online*. Accessed on April 10, 2021. https://www.hopkinsallchildrens.org/Patients-Families/Health-Library/HealthDocNew/Cutting

DARPA (2021). Reliable Neural-Interface Technology (RE-NET). DARPA agency website. Accessed on June 1, 2021. https://www.darpa.mil/program/re-net-reliable-peripheral-interfaces

Delaney, K. (2019). Battelle-led team wins DARPA award to develop injectable, bi-directional brain computer interface. Press release, via *Business Wire on AP News*, from Battelle, May 20, 2019. https://apnews.com/press-release/pr-businesswire/856d3482387741d5937af2fcec9d2314

Donnelly, G. E., Zheng, T., Haisley, E., & Norton, M. I. (2018). The amount and source of millionaires' wealth (moderately) predicts their happiness. *Personality and Social Psychology Bulletin,* May 2018. https://www.hbs.edu/faculty/Pages/item.aspx?num=53540

Duportail, J. (2017). I asked Tinder for my data: It sent me 800 pages of my deepest, darkest secrets. *The Guardian*, September 26, 2017. https://www.theguardian.com/technology/2017/sep/26/tinder-personal-data-dating-app-messages-hacked-sold

Frum, L. (2021). Bleak cyborg future from brain-computer interfaces if we're not careful. *APL Bioengineering News*, July 20, 2021. Article. American Institute of Physics. https://publishing.aip.org/publications/latest-content/bleak-cyborg-future-from-brain-computer-interfaces-if-were-not-careful/

Funnell, A. (2021). The potential of Brain-Machine Interface technology has Silicon Valley excited, and ethicists worried. Radio interview later posted as an online article on March

10, 2021. *Australian Broadcasting Corporation (ABC) Radio*. Article. https://www.abc.net.au/news/2021-03-11/the-age-of-thought-control-is-already-upon-us/13229378

Gaffary, S. (2021). Why some biologists and ecologists think social media is a risk to humanity. *Vox*, June 26, 2021. https://www.vox.com/recode/2021/6/26/22550981/carl-bergstrom-joe-bak-coleman-biologists-ecologists-social-media-risk-humanity-research-academics

Garrett, B. (2020). We should all be preppers: Chances are you have a neighbor who was ready for this pandemic. *The Atlantic*, May 3, 2020. Article. https://www.theatlantic.com/ideas/archive/2020/05/we-should-all-be-preppers/611074/

Gillespie, A. (2021). China's Social Credit System will impact religious freedoms. *Christian Today Australia*. Accessed on June 3, 2021. https://christiantoday.com.au/news/chinas-social-credit-system-will-impact-on-religious-freedoms.html

Goins-Phillips, T. (2021). Famed psychologist Jordan Peterson tears up while talking about Jesus. *CBN News Faithwire*, March 14, 2021. Article. https://cmsedit.cbn.com/cbnnews/us/2021/march/famed-psychologist-jordan-peterson-tears-up-talking-about-jesus

Gravitz, L. (2019). The importance of forgetting. *Nature*, Vol. 571, July 25, 2019, 12-14.

Greitemeyer, T., & Mügge, D. O. (2014). Video games do affect social outcomes: A meta-analytic review of the effects of violent and prosocial video game play. *Personality and Social Psychology Bulletin, 40* (5), 578-589.

"Growth and the middle class." (2011). *Democracy: A Journal of Ideas*, Spring, 2011, No. 20.

https://democracyjournal.org/magazine/20/growth-and-the-middle-class/

Hargrave, G. (2020). How to take back control of your notifications and get more done. *Wired UK*, January 1, 2020. https://www.wired.co.uk/article/control-notification

Harris, K. T. (2014). Pentagon rolls out DARPA plan to implant chip in soldiers' brains. *The Rundown Live*, February 15, 2014. Article. https://therundownlive.com/pentagon-rolls-out-darpa-plan-to-implant-chip-in-soldiers-brain/

Hastings, C. (2019). Synchron announces first successful implantation of Stentrode. *MedGadget*, September 30, 2019. https://www.medgadget.com/2019/09/synchron-announces-first-successful-clinical-implantation-of-stentrode.html

Hatmaker, T. (2021). Facebook knows Instagram harms teens. Now, its plan to open the app to kids looks worse than ever. *Tech Crunch*, September 16, 2021. Article. https://techcrunch.com/2021/09/16/facebook-instagram-for-kids-mosseri-wsj-teen-girls/

Heinlein, R. A. (1982). *Friday*. New York, NY: Holt, Rinehart, and Winston.

Herbeck, D., Brecht, M. & Lovinger, K. (2014). Mortality, causes of death, and health status among methamphetamine users. *Journal of Addictive Diseases*, *34*(1).

Herbert, F. (1985). *Chapterhouse: Dune*. New York, NY: G. P. Putnam's Sons.

Higgs, R. (1990). The growth of government in the United States. *Independent Institute News*, August 1, 1990. https://www.independent.org/news/article.asp?id=1390

Horn C. (2018). Design guide on how to miniaturize medical devices. *Machine Design*, April 24, 2018. https://www.machinedesign.com/mechanical-motion-systems/article/21836671/design-guide-on-how-to-miniaturize-medical-devices

Horowitz, J. M., Igielnik, R., & Kochhar, R. (2020). *Trends in income and wealth inequality*. Report. Washington, DC: Pew Research Center, January 9, 2020. https://www.pewresearch.org/social-trends/2020/01/09/trends-in-income-and-wealth-inequality/

Huang, C. (2020). Neurodata company Kernel looks to scale after big raise. *Los Angeles Business Journal*, July 17, 2020. https://labusinessjournal.com/news/2020/jul/09/kernel-raises-53-million-series-c/

"IBM 2nm chip breakthrough claims more power with less energy." (2021). BBC News, May 6, 2021. Article. https://www.bbc.com/news/technology-57009930

"Joe Rogan Experience #1188." (2018). October 24, 2018. Podcast. https://www.youtube.com/watch?v=j5FOumrXyww

Kearns, C. E., Schmidt, L. A., Glantz, S. A. (2016). Sugar industry and coronary heart disease research: A historical analysis of internal industry documents. *JAMA Internal Medicine*, 176(11): 1680-1685.

Kenneally, C. (2021). Do brain implants change your identity? *The New Yorker*, The Technology Issue, April 26 & May 3, 2021.

Kharpal, A. (2021). In battle with U.S., China to focus on 7 "frontier" technologies from chips to brain-computer fusion. *CNBC*, March 5, 2021. Article. https://www.cnbc.com/2021/03/05/china-to-focus-on-frontier-tech-from-chips-to-quantum-computing.html

Kharpal, A. (2017). Stephen Hawking says AI could be "worst event in the history of our civilization." *CNBC*, November 6, 2017. Article. https://www.cnbc.com/2017/11/06/stephen-hawking-ai-could-be-worst-event-in-civilization.html

Kim, H. (2013). Exercise rehabilitation for smartphone addiction. *Journal of Exercise Rehabilitation*, *9*(6).

Klein, N. (2007). *The shock doctrine: The rise of disaster capitalism*. Toronto: Random House of Canada.

Kurutz, S. (2021). Suspicious minds. *The New York Times*, July 21, 2018. https://www.nytimes.com/2018/07/21/style/ancient-aliens.html

Labonte, M. (2010). The size and role of government: Economic issues. *CRS Report for Congress*, June 14, 2010. Washington, DC: Congressional Research Service. https://fas.org/sgp/crs/misc/RL32162.pdf

Leslie, I. (2016). The sugar conspiracy. *The Guardian*, April 7, 2016. https://www.theguardian.com/society/2016/apr/07/the-sugar-conspiracy-robert-lustig-john-yudkin

Lewis, C. S. (1942). *The Screwtape letters*. New York, NY: HarperCollins.

Lukianoff, G., & Haidt., J. (2018). *The coddling of the American mind: How good intentions and bad ideas are setting up a generation for failure*. New York, NY: Penguin Random House.

MacIntyre, I. (2013). *The chronology of major eschatological events in chapters 4 to 19 of Revelation.* Master's thesis. Charlotte, NC: Reformed Theological Seminary.

Markoff, J. (2019). Elon Musk's Neuralink wants "sewing machine like" robots to wire brains to the Internet. *The New York Times*, July 16, 2019. https://www.nytimes.com/2019/07/16/technology/neuralink-elon-musk.html?searchResultPosition=1

McBride, S. (2021). Neuralink competitor raises $20 million for brain implants. *Bloomberg News*, July 22, 2021. Article. https://www.bloomberg.com/news/articles/2021-07-22/neuralink-competitor-raises-20-million-for-brain-implants

Milgram, S. (1974). *Obedience to authority: An experimental view*. New York, NY: Harper & Row.

Mitnick, K., & Vamosi, R. (2017). *The art of invisibility: The world's most famous hacker teaches you how to be safe in the age of Big Brother and big data*. New York, NY: Back Bay Books.

Moore, T. J., Glenmullen, J., & Mattison, D. R. (2014). Reports of pathological gambling, hypersexuality, and compulsive shopping associated with dopamine receptor agonist drugs. *JAMA Internal Medicine, 174*(12), pp. 1930-1933.

Mosseri, A. (2021). Pausing "Instagram Kids" and building parental supervision tools. Instagram, September 27, 2021. https://about.instagram.com/blog/announcements/pausing-instagram-kids

OECD. (2019). *Under pressure: The squeezed middle class*. Paris: OECD Publishing.

Parnell, B. (2017). Is social media hurting your mental health? *TEDx Talks*. Video. https://www.youtube.com/watch?v=Czg_9C7gw0o

Pinsker, J. (2018). The reason many ultrarich people aren't satisfied with their wealth. *The Atlantic,* December 4, 2018. https://www.theatlantic.com/family/archive/2018/12/rich-people-happy-money/577231/

Plumhoff, K. (2021). Attack of the phones. *Health Magazine Special Edition: Understanding Addiction*. New York, NY: Meredith.

Portillo-Lara, R. et al. (2021). Mind the gap: State-of-the-art technologies and applications for EEG-based brain-computer interfaces. *APL Bioengineering*, 5(3). https://aip.scitation.org/doi/10.1063/5.0047237

"Q Who?" (1989). *Star Trek: The Next Generation*, Season 2, Episode 16. Originally aired on May 5, 1989. Television show. Paramount Domestic Television.

Quiroga, R. Q. (2019). "Funes the Memorious" and other cases of extraordinary memory. *The MIT Press Reader Online.* Cambridge, MA: The MIT Press. https://thereader.mitpress.mit.edu/borges-memory-funes-the-memorious/

Scott, C. (2014). Parkinson's drugs may lead to compulsive gambling, shopping, and sex. *Healthline*, October 20, 2014. Article. https://www.healthline.com/health-news/parkinsons-drugs-may-lead-to-compulsive-gambling-102014

Shahar, D. & Sayers, M. G. L. (2018). Prominent exostosis projecting from the occipital squama more substantial and prevalent in young adults than older age groups. *Scientific Reports*, *8*, Article 3354.

Spichak, S. (2021). Departed Neuralink co-founder locks down $47 million for secretive neuroscience startup. *Futurism*, August 24, 2021. Article. https://futurism.com/neuralink-max-hodak-secretive-startup

SureCall Company. (2018). *The attachment problem: Cellphone use in America*. Survey. https://www.surecall.com/docs/20180515-SureCall-Attachment-Survey-Results-v2.pdf

"The Game." (1991). *Star Trek: The Next Generation*, Season 5, Episode 6. Originally aired on October 28, 1991. Television show. Paramount Domestic Television.

The Institute of Arts and Sciences. (2012) Artificial Intelligence and consciousness: David Malone. March 3, 2020. Video interview. https://iai.tv/video/david-malone-in-depth-interview-dualism-consciousness?utm_source =YouTube&utm_medium=description

"This hot robot says she wants to destroy humans." (2016). *CNBC's The Pulse*, March 16, 2016. Video. https://www.cnbc.com/video/2016/03/16/this-hot-robot-says-she-wants-to-destroy-humans.html?&qsearchterm =robot%20destroy%20all%20humans

Toffler, A. (1970). *Future shock*. New York, NY: Random House.

Tregelles, S. P. (1849). *The book of Revelation: Translated from the ancient Greek text*. English translation. London: Samuel Bagster & Sons.

Tsotsis, A. (2013). Employer tipped off police to pressure cooker and backpack searches, not Google. *Tech Crunch*, August 1, 2013. Article. https://techcrunch.com/2013/08/01/employer-tipped-off-police-in-pressure-cookerbackpack-gate-not-google/

Twenge, J. M. (2017). *iGen: Why today's super-connected kids are growing up less rebellious, more tolerant, less happy—and completely unprepared for adulthood—and what that means for the rest of us*. New York, NY: Atria.

United States Census Bureau. (2010). What percent do public employees make up of total employees in the United States? https://www.census.gov/newsroom/cspan/households_and_businesses/20120720_cspan_hh_bus_slides_13.pdf

Van Doren, C. (1991). *A history of knowledge: Past, present, and future*. New York, NY: Citadel Press.

"Watch Elon Musk's entire live Neuralink demonstration." (2020). *CNET*, August 28, 2020. YouTube stream. https://www.youtube.com/watch?v=iOWFXqT5MZ4&t=399s

Wells, G., Horwitz, J., & Seetharaman, D. (2021). Facebook knows Instagram is toxic for teen girls, company documents show. *The Wall Street Journal*, September 14, 2021. https://www.wsj.com/articles/facebook-knows-instagram-is-toxic-for-teen-girls-company-documents-show-11631620739

Winnick, M. (2016). *Mobile touches: Inaugural study on humans and their tech*. Chicago, IL: Dscout.

Witze, A. (2020). Collective behavior: How animals work together. *Knowable Magazine*, July 29, 2020. https://knowablemagazine.org/article/living-world/2020/collective-behavior-how-animals-work-together

"You Already Carry the Mark of the Beast. Don't Believe? You Will Soon." (2014). *Catholic Online*, April 28, 2014. https://www.catholic.org/news/technology/story.php?id=55178

APPENDIX A: LETTERS TO CONSIDER WRITING

Consider writing to your U.S. representative, your two U.S. senators, and the president. Let them know that you oppose brain-tech implants for mass production and mass insertion in human skulls, and that you support federal legislation banning them.

Letter to Your Representative in the U.S. House of Representatives

Your Full Name
Address
City, State and Zip Code

[Insert the date]

The Honorable [Insert your representative's full name]
U.S. House of Representatives
Washington, DC 20515

Dear Representative [Insert your representative's last name]:

My name is [Insert your full name] and I reside in your Congressional district.

I am writing to you regarding the issue of brain-machine interface technology, and the upcoming plans by some corporations to have brain-tech implants mass produced and mass marketed to the general public. My suggestion is that the U.S. House of Representatives consider passing legislation that bans these types of imbedded devices.

The implants are being hyped as providing the capability for instant learning; instant memory; cures for some neurological disorders; and calling up the Web in one's mind. However, upon further inspection, they strongly resemble smartphones, each with a Bluetooth, a processor, and a wireless charger. Brainphones, so to speak.

Smartphones, along with the social media they facilitate, are currently causing dramatic psychological and physiological harm to humans, especially young people. Dr. Jean Twenge's research at San Diego State University, as one example, is compelling, tying the devices to an increase in suicide, poor sleep habits, and declining health. Smartphones and social media also tend to be highly addictive. With implants, the tendency for humans to always be "on" and always be connected is frightening. The stress and trauma currently brought on by smartphone use might increase dramatically.

The United States could lead the world in protecting its citizens, particularly its young people, from these devices. Strong legislation banning the implants would make a big difference.

Thank you for considering my request. And thank you for your public service to this great nation.

Sincerely,
[Insert your full name]

Letter to One of Your Two Senators in the U.S. Senate

Your Full Name
Address
City, State and Zip Code

[Insert the date]

The Honorable [Insert your senator's full name]
United States Senate
Washington, D.C. 20510

Dear Senator [Insert your senator's last name]:

My name is [Insert your full name] and I reside in [Insert your state].

I am writing to you regarding the issue of brain-machine interface technology, and the upcoming plans by some corporations to have brain-tech implants mass produced and mass marketed to the general public. My suggestion is that the U.S. Senate consider passing legislation that bans these types of imbedded devices.

The implants are being hyped as providing the capability for instant learning; instant memory; cures for some neurological disorders; and calling up the Web in one's mind. However, upon further inspection, they strongly resemble smartphones, each with a Bluetooth, a processor, and a wireless charger. Brainphones, so to speak.

Smartphones, along with the social media they facilitate, are currently causing dramatic psychological and physiological harm to humans, especially young people. Dr. Jean Twenge's research at San Diego State University, as one example, is compelling, tying the devices to an increase in suicide, poor sleep habits, and declining health. Smartphones and social media also tend to be highly addictive. With implants, the tendency for humans to always be "on" and always be connected is frightening. The stress and trauma currently brought on by smartphone use might increase dramatically.

The United States could lead the world in protecting its citizens, particularly its young people, from these devices. Strong legislation banning the implants would make a big difference.

Thank you for considering my request. And thank you for your public service to this great nation.

Sincerely,
[Insert your name]

Letter to the President of the United States

Your Name
Address
City, State, Zip Code

[Insert the date]

The President
The White House
1600 Pennsylvania Avenue, N.W.
Washington, DC 20500

Dear Mr. [or Madam] President:

My name is [Insert your full name] and I reside in [Insert your state].

I am writing to you regarding the issue of brain-machine interface technology, and the upcoming plans by some corporations to have brain-tech implants mass produced and mass marketed to the general public. My suggestion is that you consider urging legislation that bans these types of imbedded devices.

The implants are being hyped as providing the capability for instant learning; instant memory; cures for some neurological disorders; and calling up the Web in one's mind. However, upon further inspection, they strongly resemble smartphones, each with a Bluetooth, a processor, and a wireless charger. Brainphones, so to speak.

Smartphones, along with the social media they facilitate, are currently causing dramatic psychological and physiological harm to humans, especially young people. Dr. Jean Twenge's research at San Diego State University, as one example, is compelling, tying the devices to an increase in suicide, poor sleep habits, and declining health. Smartphones and social media also tend to be highly addictive. With implants, the tendency for humans to always be "on" and always be connected is frightening. The stress and trauma currently brought on by smartphone use might increase dramatically.

The United States could lead the world in protecting its citizens, particularly its young people, from these devices. Strong legislation banning the implants would make a big difference.

Thank you for considering my request. And thank you for your service to this great nation.

Sincerely,
[Insert your full name]

APPENDIX B:
WAYS TO STOP TECHNOLOGY ADDICTION

Consider the following steps if the craving for technology is making your life or the lives of your family members unmanageable. Breaking your current dependence on technology will make saying "no" to the brainphone much easier. These steps are drawn, in part, from Buttimer (2021), Plumhoff (2021), and Hargrave (2020).

- Talk to your doctor, your professional counselor, or your minister about your technology addiction. You won't be the first person who has done so—maybe not even the first person today.
- Set a good example for your family. Maintain a log of how much time you spend on your technology. Work towards reducing the time. Remember, the mental and physical health of you and your family depends on it.
- Turn off all the notifications and app prompts on your smartphone. Be ruthless.
- Uninstall the social media apps from your smartphone. That is, limit your social media exposure to your time on a laptop or tablet.
- Shut off your smartphone, laptop, tablet, virtual helmet, etc., entirely for 60 minutes before bedtime. Doing so will allow your brain to decompress from all that screen stimulation, making it easier to fall asleep.
- Keep your smartphone off while you sleep. Purchase an old-fashioned alarm clock to wake you up in the morning.
- Set aside regular, technology-free times each day for you and your family. Meals together, without smartphones or virtual helmets, are a nice starting point.

- If you have the time, the inclination, and the physical ability, take up a non-virtual sport, such as hiking. If it is a competitive sport, such as bowling, join a team.
- Remember, you're in charge of your kids when it comes to technology restrictions and time limitations—not the other way around. If your child is gaming on his computer or virtual helmet at two in the morning and urinating into a plastic bottle because he doesn't want to lose game points by leaving the screen, there's a problem.
- Spend time each day completing a small list of things requiring no technology. Make your bed. Keep a written journal. Have a face-to-face conversation with someone. Go for a walk. Read three pages from a paper book.
- Join a Twelve-Step group. It doesn't have to be a technology group, and you're not obligated to tell anyone in the group what addiction you're there for. For example, you can join an Alcoholics Anonymous, Narcotics Anonymous, or Overeaters Anonymous group, and follow along with the steps as they pertain to your addiction to technology.
- Don't get a brainphone or any other type of technology implant that allows the Internet or gaming straight into your brain. It is next to impossible to cut back on something you're connected to 24 hours a day. Again, the mental and physical health of you and your family depends on it.

INDEX

OBELISKS: TOWERS OF POWER
The Mysterious Purpose of Obelisks
By David Hatcher Childress

Some obelisks weigh over 500 tons and are massive blocks of polished granite that would be extremely difficult to quarry and erect even with modern equipment. Why did ancient civilizations in Egypt, Ethiopia and elsewhere undertake the massive enterprise it would have been to erect a single obelisk, much less dozens of them? Were they energy towers that could receive or transmit energy? Chapters include: Megaliths Around the World and their Purpose; The Crystal Towers of Egypt; The Obelisks of Ethiopia; Obelisks in Europe and Asia; Mysterious Obelisks in the Americas; The Terrible Crystal Towers of Atlantis; Tesla's Wireless Power Distribution System; Obelisks on the Moon; more. 8-page color section.

336 Pages. 6x9 Paperback. Illustrated. $22.00 Code: OBK

MEN & GODS IN MONGOLIA
by Henning Haslund

Haslund takes us to the lost city of Karakota in the Gobi desert. We meet the Bodgo Gegen, a god-king in Mongolia similar to the Dalai Lama of Tibet. We meet Dambin Jansang, the dreaded warlord of the "Black Gobi." Haslund and companions journey across the Gobi desert by camel caravan; are kidnapped and held for ransom; withness initiation into Shamanic societies; meet reincarnated warlords; and experience the violent birth of "modern" Mongolia.

358 Pages. 6x9 Paperback. Illustrated. $18.95. Code: MGM

PROJECT MK-ULTRA AND MIND CONTROL TECHNOLOGY
By Axel Balthazar

This book is a compilation of the government's documentation on MK-Ultra, the CIA's mind control experimentation on unwitting human subjects, as well as over 150 patents pertaining to artificial telepathy (voice-to-skull technology), behavior modification through radio frequencies, directed energy weapons, electronic monitoring, implantable nanotechnology, brain wave manipulation, nervous system manipulation, neuroweapons, psychological warfare, satellite terrorism, subliminal messaging, and more. A must-have reference guide for targeted individuals and anyone interested in the subject of mind control technology.

384 pages. 7x10 Paperback. Illustrated. $19.95. Code: PMK

LIQUID CONSPIRACY 2:
The CIA, MI6 & Big Pharma's War on Psychedelics
By Xaviant Haze

Underground author Xaviant Haze looks into the CIA and its use of LSD as a mind control drug; at one point every CIA officer had to take the drug and endure mind control tests and interrogations to see if the drug worked as a "truth serum." Chapters include: The Pioneers of Psychedelia; The United Kingdom Mellows Out: The MI5, MDMA and LSD; Taking it to the Streets: LSD becomes Acid; Great Works of Art Inspired and Influenced by Acid; Scapolamine: The CIA's Ultimate Truth Serum; Mind Control, the Death of Music and the Meltdown of the Masses; Big Pharma's War on Psychedelics; The Healing Powers of Psychedelic Medicine; tons more.

240 pages. 6x9 Paperback. Illustrated. $19.95. Code: LQC2

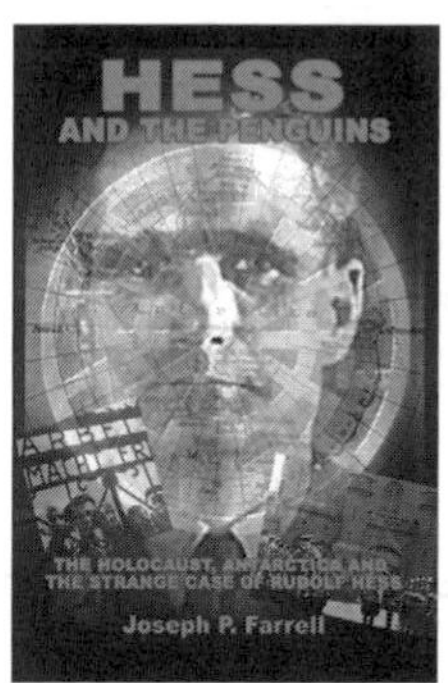

HESS AND THE PENGUINS
The Holocaust, Antarctica and the Strange Case of Rudolf Hess
By Joseph P. Farrell

Farrell looks at Hess' mission to make peace with Britain and get rid of Hitler—even a plot to fly Hitler to Britain for capture! How much did Göring and Hitler know of Rudolf Hess' subversive plot, and what happened to Hess? Why was a doppleganger put in Spandau Prison and then "suicided"? Did the British use an early form of mind control on Hess' double? John Foster Dulles of the OSS and CIA suspected as much. Farrell also uncovers the strange death of Admiral Richard Byrd's son in 1988, about the same time of the death of Hess.

288 Pages. 6x9 Paperback. Illustrated. $19.95. Code: HAPG

HIDDEN FINANCE, ROGUE NETWORKS & SECRET SORCERY
The Fascist International, 9/11, & Penetrated Operations
By Joseph P. Farrell

Farrell investigates the theory that there were not *two* levels to the 9/11 event, but *three*. He says that the twin towers were downed by the force of an exotic energy weapon, one similar to the Tesla energy weapon suggested by Dr. Judy Wood, and ties together the tangled web of missing money, secret technology and involvement of portions of the Saudi royal family. Farrell unravels the many layers behind the 9-11 attack, layers that include the Deutschebank, the Bush family, the German industrialist Carl Duisberg, Saudi Arabian princes and the energy weapons developed by Tesla before WWII.

296 Pages. 6x9 Paperback. Illustrated. $19.95. Code: HFRN

THRICE GREAT HERMETICA & THE JANUS AGE
By Joseph P. Farrell

What do the Fourth Crusade, the exploration of the New World, secret excavations of the Holy Land, and the pontificate of Innocent the Third all have in common? Answer: Venice and the Templars. What do they have in common with Jesus, Gottfried Leibniz, Sir Isaac Newton, Rene Descartes, and the Earl of Oxford? Answer: Egypt and a body of doctrine known as Hermeticism. The hidden role of Venice and Hermeticism reached far and wide, into the plays of Shakespeare (a.k.a. Edward DeVere, Earl of Oxford), into the quest of the three great mathematicians of the Early Enlightenment for a lost form of analysis, and back into the end of the classical era, to little known Egyptian influences at work during the time of Jesus.

354 Pages. 6x9 Paperback. Illustrated. $19.95. Code: TGHJ

ROBOT ZOMBIES
Transhumanism and the Robot Revolution
By Xaviant Haze and Estrella Eguino,

Technology is growing exponentially and the moment when it merges with the human mind, called "The Singularity," is visible in our imminent future. Science and technology are pushing forward, transforming life as we know it—perhaps even giving humans a shot at immortality. Who will benefit from this? This book examines the history and future of robotics, artificial intelligence, zombies and a Transhumanist utopia/dystopia integrating man with machine. Chapters include: Love, Sex and Compassion—Android Style; Humans Aren't Working Like They Used To; Skynet Rises; Blueprints for Transhumans; Kurzweil's Quest; Nanotech Dreams; Zombies Among Us; Cyborgs (Cylons) in Space; Awakening the Human; more. Color Section.

180 Pages. 6x9 Paperback. Illustrated. $16.95. Code: RBTZ

THE LOST WORLD OF CHAM
The Trans-Pacific Voyages of the Champa
By David Hatcher Childress

The mysterious Cham, or Champa, peoples of Southeast Asia formed a megalith-building, seagoing empire that extended into Indonesia, Tonga, and beyond—a transoceanic power that reached Mexico and South America. The Champa maintained many ports in what is today Vietnam, Cambodia, and Indonesia and their ships plied the Indian Ocean and the Pacific, bringing Chinese, African and Indian traders to far off lands, including Olmec ports on the Pacific Coast of Central America. Topics include: Cham and Khem: Egyptian Influence on Cham; The Search for Metals; The Basalt City of Nan Madol; Elephants and Buddhists in North America; The Cham and Lake Titicaca; Easter Island and the Cham; the Magical Technology of the Cham; tons more. 24-page color section.

328 Pages. 6x9 Paperback. Illustrated. $22.00 Code: LPWC

ADVENTURES OF A HASHISH SMUGGLER
by Henri de Monfreid

Nobleman, writer, adventurer and inspiration for the swashbuckling gun runner in the *Adventures of Tintin*, Henri de Monfreid lived by his own account "a rich, restless, magnificent life" as one of the great travelers of his or any age. The son of a French artist who knew Paul Gaugin as a child, de Monfreid sought his fortune by becoming a collector and merchant of the fabled Persian Gulf pearls. He was then drawn into the shadowy world of arms trading, slavery, smuggling and drugs. Infamous as well as famous, his name is inextricably linked to the Red Sea and the raffish ports between Suez and Aden in the early years of the twentieth century. De Monfreid (1879 to 1974) had a long life of many adventures around the Horn of Africa where he dodged pirates as well as the authorities.

284 Pages. 6x9 Paperback. $16.95. Illustrated. Code AHS

NORTH CAUCASUS DOLMENS
In Search of Wonders
By Boris Loza, Ph.D.

Join Boris Loza as he travels to his ancestral homeland to uncover and explore dolmens firsthand. Throughout this journey, you will discover the often hidden, and surprisingly forbidden, perspective about the mysterious dolmens: their ancient powers of fertility, healing and spiritual connection. Chapters include: Ancient Mystic Megaliths; Who Built the Dolmens?; Why the Dolmens were Built; Asian Connection; Indian Connection; Greek Connection; Olmec and Maya Connection; Sun Worshippers; Dolmens and Archeoastronomy; Location of Dolmen Quarries; Hidden Power of Dolmens; and much more! Tons of Illustrations! A fascinating book of little-seen megaliths. Color section.

252 Pages. 5x9 Paperback. Illustrated. $24.00. Code NCD

GIANTS: MEN OF RENOWN
By Denver Michaels

Michaels runs down the many stories of giants around the world and testifies to the reality of their existence in the past. Chapters and subchapters on: Giants in the Bible; Texts; Tales from the Maya; Stories from the South Pacific; Giants of Ancient America; The Stonish Giants; Mescalero Tales; The Nahullo; Mastodons, Mammoths & Mound Builders; Pawnee Giants; The Si-Te-Cah; Tsul 'Kalu; The Titans & Olympians; The Hyperboreans; European Myths; The Giants of Britain & Ireland; Norse Giants; Myths from the Indian Subcontinent; Daityas, Rakshasas, & More; Jainism: Giants & Inconceivable Lifespans; The Conquistadors Meet the Sons of Anak; Cliff-Dwelling Giants; The Giants of the Channel Islands; Strange Tablets & Other Artifacts; more. Tons of illustrations with an 8-page color section.

320 Pages. 6x9 Paperback. Illustrated. $22.00. Code: GMOR

HAUNEBU: THE SECRET FILES
The Greatest UFO Secret of All Time
By David Hatcher Childress

Childress brings us the incredible tale of the German flying disk known as the Haunebu. Although rumors of German flying disks have been around since the late years of WWII it was not until 1989 when a German researcher named Ralf Ettl living in London received an anonymous packet of photographs and documents concerning the planning and development of at least three types of unusual craft. Chapters include: A Saucer Full of Secrets; WWII as an Oil War; A Saucer Called Vril; Secret Cities of the Black Sun; The Strange World of Miguel Serrano; Set the Controls for the Heart of the Sun; Dark Side of the Moon: more. Includes a 16-page color section. Over 120 photographs and diagrams.

352 Pages. 6x9 Paperback. Illustrated. $22.00 Code: HBU

HIDDEN AGENDA
NASA and the Secret Space Program
By Mike Bara

Bara delves into secret bases on the Moon, and exploring the many other rumors surrounding the military's secret projects in space. On June 8, 1959, a group at the ABMA produced for the US Department of the Army a report entitled Project Horizon, a "Study for the Establishment of a Lunar Military Outpost." The permanent outpost was predicted to cost $6 billion and was to become operational in December 1966 with twelve soldiers stationed at the Moon base. Does hacker Gary Mackinnon's discovery of defense department documents identifying "non-terrestrial officers" serving in space? Includes an 8-page color section.

346 Pages. 6x9 Paperback. Illustrated. $19.95. Code: HDAG

THE ANTI-GRAVITY FILES
A Compilation of Patents and Reports
Edited by David Hatcher Childress

With plenty of technical drawings and explanations, this book reveals suppressed technology that will change the world in ways we can only dream of. Chapters include: A Brief History of Anti-Gravity Patents; The Motionless Electromagnet Generator Patent; Mercury Anti-Gravity Gyros; The Tesla Pyramid Engine; Anti-Gravity Propulsion Dynamics; The Machines in Flight; More Anti-Gravity Patents; Death Rays Anyone?; The Unified Field Theory of Gravity; and tons more. Heavily illustrated. 4-page color section.

216 pages. 8x10 Paperback. Illustrated. $22.00. Code: AGF

SECRET MARS: The Alien Connection
By M. J. Craig

While scientists spend billions of dollars confirming that microbes live in the Martian soil, people sitting at home on their computers studying the Mars images are making far more astounding discoveries… they have found the possible archaeological remains of an extraterrestrial civilization. Hard to believe? Well, this challenging book invites you to take a look at the astounding pictures yourself and make up your own mind. *Secret Mars* presents over 160 incredible images taken by American and European spacecraft that reveal possible evidence of a civilization that once lived, and may still live, on the planet Mars… powerful evidence that scientists are ignoring! A visual and fascinating book!

352 Pages. 6x9 Paperback. Illustrated. $19.95. Code: SMAR

SAUCERS, SWASTIKAS AND PSYOPS
A History of a Breakaway Civilization
By Joseph P. Farrell

Farrell discusses SS Commando Otto Skorzeny; George Adamski; the alleged Hannebu and Vril craft of the Third Reich; The Strange Case of Dr. Hermann Oberth; Nazis in the US and their connections to "UFO contactees"; The Memes—an idea or behavior spread from person to person within a culture—are Implants. Chapters include: The Nov. 20, 1952 Contact: The Memes are Implants; The Interplanetary Federation of Brotherhood; Adamski's Technological Descriptions and Another ET Message: The Danger of Weaponized Gravity; Adamski's Retro-Looking Saucers, and the Nazi Saucer Myth; Dr. Oberth's 1968 Statements on UFOs and Extraterrestrials; more.

272 Pages. 6x9 Paperback. Illustrated. $19.95. Code: SSPY

LBJ AND THE CONSPIRACY TO KILL KENNEDY
By Joseph P. Farrell

Farrell says that a coalescence of interests in the military industrial complex, the CIA, and Lyndon Baines Johnson's powerful and corrupt political machine in Texas led to the events culminating in the assassination of JFK. Chapters include: Oswald, the FBI, and the CIA: Hoover's Concern of a Second Oswald; Oswald and the Anti-Castro Cubans; The Mafia; Hoover, Johnson, and the Mob; The FBI, the Secret Service, Hoover, and Johnson; The CIA and "Murder Incorporated"; Ruby's Bizarre Behavior; The French Connection and Permindex; Big Oil; The Dead Witnesses: Guy Bannister, Jr., Mary Pinchot Meyer, Rose Cheramie, Dorothy Killgallen, Congressman Hale Boggs; LBJ and the Planning of the Texas Trip; LBJ: A Study in Character, Connections, and Cabals; LBJ and the Aftermath: Accessory After the Fact; The Requirements of Coups D'État; more.

342 Pages. 6x9 Paperback. $19.95 Code: LCKK

THE TESLA PAPERS
Nikola Tesla on Free Energy & Wireless Transmission of Power
by Nikola Tesla, edited by David Hatcher Childress

David Hatcher Childress takes us into the incredible world of Nikola Tesla and his amazing inventions. Tesla's fantastic vision of the future, including wireless power, anti-gravity, free energy and highly advanced solar power. Also included are some of the papers, patents and material collected on Tesla at the Colorado Springs Tesla Symposiums, including papers on: •The Secret History of Wireless Transmission •Tesla and the Magnifying Transmitter •Design and Construction of a Half-Wave Tesla Coil •Electrostatics: A Key to Free Energy •Progress in Zero-Point Energy Research •Electromagnetic Energy from Antennas to Atoms

325 PAGES. 8X10 PAPERBACK. ILLUSTRATED. $16.95. CODE: TTP

COVERT WARS & THE CLASH OF CIVILIZATIONS
UFOs, Oligarchs and Space Secrecy
By Joseph P. Farrell

Farrell's customary meticulous research and sharp analysis blow the lid off of a worldwide web of nefarious financial and technological control that very few people even suspect exists. He elaborates on the advanced technology that they took with them at the "end" of World War II and shows how the breakaway civilizations have created a huge system of hidden finance with the involvement of various banks and financial institutions around the world. He investigates the current space secrecy that involves UFOs, suppressed technologies and the hidden oligarchs who control planet earth for their own gain and profit.

358 Pages. 6x9 Paperback. Illustrated. $19.95. Code: CWCC

ORDER FORM

10% Discount When You Order 3 or More Items!

One Adventure Place
P.O. Box 74
Kempton, Illinois 60946
United States of America
Tel.: 815-253-6390 • Fax: 815-253-6300
Email: auphq@frontiernet.net
http://www.adventuresunlimitedpress.com

ORDERING INSTRUCTIONS

- ✓ Remit by USD$ Check, Money Order or Credit Card
- ✓ Visa, Master Card, Discover & AmEx Accepted
- ✓ Paypal Payments Can Be Made To: info@wexclub.com
- ✓ Prices May Change Without Notice
- ✓ 10% Discount for 3 or More Items

SHIPPING CHARGES

United States

- ✓ POSTAL BOOK RATE
- ✓ Postal Book Rate { $4.50 First Item; 50¢ Each Additional Item
- ✓ Priority Mail { $7.00 First Item; $2.00 Each Additional Item
- ✓ UPS { $9.00 First Item (Minimum 5 Books); $1.50 Each Additional Item

NOTE: UPS Delivery Available to Mainland USA Only

Canada

- ✓ Postal Air Mail { $19.00 First Item; $3.00 Each Additional Item
- ✓ Personal Checks or Bank Drafts MUST BE US$ and Drawn on a US Bank
- ✓ Canadian Postal Money Orders OK
- ✓ Payment MUST BE US$

All Other Countries

- ✓ Sorry, No Surface Delivery!
- ✓ Postal Air Mail { $19.00 First Item; $7.00 Each Additional Item
- ✓ Checks and Money Orders MUST BE US$ and Drawn on a US Bank or branch.
- ✓ Paypal Payments Can Be Made in US$ To: info@wexclub.com

SPECIAL NOTES

- ✓ RETAILERS: Standard Discounts Available
- ✓ BACKORDERS: We Backorder all Out-of-Stock Items Unless Otherwise Requested
- ✓ PRO FORMA INVOICES: Available on Request
- ✓ DVD Return Policy: Replace defective DVDs only

ORDER ONLINE AT: www.adventuresunlimitedpress.com

10% Discount When You Order 3 or More Items!

Please check: ✓

☐ This is my first order ☐ I have ordered before

Name

Address

City

State/Province | Postal Code

Country

Phone: Day | Evening

Fax | Email

Item Code	Item Description	Qty	Total

Please check: ✓

- ☐ Postal-Surface
- ☐ Postal-Air Mail (Priority in USA)
- ☐ UPS (Mainland USA only)

Subtotal ▶	
Less Discount-10% for 3 or more items ▶	
Balance ▶	
Illinois Residents 6.25% Sales Tax ▶	
Previous Credit ▶	
Shipping ▶	
Total (check/MO in USD$ only) ▶	

☐ Visa/MasterCard/Discover/American Express

Card Number:

Expiration Date: | Security Code:

✓ SEND A CATALOG TO A FRIEND: